Climate and Society in Colonial Mexico

T0260994

RGS-IBG Book Series

Climate and Society in Colonial Mexico

A Study in Vulnerability

Georgina H. Endfield

BLACKWELL PUBLISHING
350 Main Street, Malden, MA 02148–5020, USA
9600 Garsington Road, Oxford OX4 2DQ, UK
550 Swanston Street, Carlton, Victoria 3053, Australia

First published 2008 by Blackwell Publishing Ltd

1   2008

Library of Congress Cataloging-in-Publication Data

Endfield, Georgina H.
      Climate and society in colonial Mexico : a study in vulnerability / Georgina H. Endfield.
         p. cm. – (RGS-IBG book series)
      Includes bibliographical references and index.
      ISBN 978-1-4051-4582-4 (pbk. : alk. paper) – ISBN 978-1-4051-4583-1 (hardcover : alk. paper)
1. Human ecology–Mexico–History. 2. Water resources–Mexico–History. 3. Subsistence economy–Mexico–History. 4. Mexico–Environmental conditions. 5. Mexico–History–Spanish colony, 1540–1810. I. Title.

      GF516.E63 2008
      304.2'50972–dc22

                                   2007026741

A catalogue record for this title is available from the British Library.

Set in 10/12pt Plantin
by SPi Publisher Services, Pondicherry, India
Printed and bound in Singapore
by Utopia Press Pte Ltd

For further information on
Blackwell Publishing, visit our website at
www.blackwellpublishing.com

Contents

List of Tables and Figures

Tables

Figures

Series Editors' Preface

The RGS-IBG Book Series only publishes work of the highest international standing. Its emphasis is on distinctive new developments in human and physical geography, although it is also open to contributions from cognate disciplines whose interests overlap with those of geographers. The Series places strong emphasis on theoretically-informed and empirically-strong texts. Reflecting the vibrant and diverse theoretical and empirical agendas that characterize the contemporary discipline, contributions are expected to inform, challenge and stimulate the reader. Overall, the RGS-IBG Book Series seeks to promote scholarly publications that leave an intellectual mark and change the way readers think about particular issues, methods or theories.

For details on how to submit a proposal please visit:
www.blackwellpublishing.com/pdf/rgsibg.pdf

Kevin Ward
University of Manchester, UK

Joanna Bullard
Loughborough University, UK

RGS-IBG Book Series Editors

Acknowledgements

The research for this volume was made possible by a research grant from the Arts and Humanities Research Council (B/RG/AN6160/APN10797) and a Philip Leverhulme Prize. The study would not have come to fruition without the help, advice and encouragement of a number of colleagues, friends and of course, my family.

I have been fortunate in working with Sarah O'Hara (who brought the Archivo General de la Nación and the regional archives of Mexico to my attention), first as an academic supervisor and latterly as a colleague and friend. I have also had the pleasure and privilege of working with Dra Isabel Fernández Tejedo over the last few years. Isabel's advice and knowledge of Mexican history have been pivotal to the completion of this study.

I would like to acknowledge the tireless assistance and patience of the porteros in Galleria 4 of the Archivo General de la Nación during my regular visits over the last thirteen years, and for putting up with endless requests for volumes of Tierras and Mercedes, day after day. I would also like to express my gratitude to unique individuals like Don Luis de Castañeda for opening up his own private archive (and the front room of his house!) to myself and Isabel while we were working in Oaxaca in 2001.

The encouragement of a number of colleagues here in Nottingham, not least Mike Heffernan and Charles Watkins, provided me with the stimulus and confidence to embark on this study, while Chris Lewis kindly lent his cartographic skills to the production of the maps.

I particularly wish to thank my husband Stephen, and my family, Mum, Dad, brothers Richard and Jason and my Grandma, for all their encouragement, enthusiasm, patience and sense of humour.

Chapter One

A Vulnerable Society

Introduction

28 June 1692 was a very wet day in Celaya, Guanajuato. Unusually heavy rains began falling late in the afternoon and continued all through the evening, causing a rapid increase in the level of the River Laja which ran adjacent to the town. Finally, just after 10 o'clock that night, the river burst its banks. The flood caused 'terrible panic among all the inhabitants of the city' and took place when the people of Celaya were at their most vulnerable.

> With the darkness of the night no one knew where the most danger lay and much less how to escape this ... men, women and children were desperately shouting and crying ... and the turbulent waters invaded everything with impetuous force ... no one believed there was any possible escape from this and all began begging for mercy.

Daylight revealed the scale of the devastation. 'The whole city was left like an immense field covered in mud, uprooted trees and branches, rubbish, debris, bodies of all kinds of animals and ... all the other remnants of the near destruction.' Volunteers from local communities joined forces with army troops to clear the debris in search of flood victims, some of whom had died trapped in their houses. In total it was estimated that close to three thousand families lost their houses and livelihoods (Zamarroni Arroyo, 1960: 141; Marmolejo, 1967).

But this calamitous event could not have come at a worse time. Repeated droughts in 1690 and 1691, followed by crop blights, had already led to harvest losses across a broad area. Famine and epidemic disease had then gripped this, and other, regions of the country, causing massive life loss, especially among the disenfranchised indigenous population, some of whom were left without access to even the most basic provisions (Orozco y Berra, 1938; Berthe, 1970). The subsistence crisis that followed resulted in social unrest and the emergence of popular uprisings across central Mexico.

Dramatic as this period was, it was in no way unique. Extreme and unusual weather events and natural hazards, including earthquakes and volcanic eruptions, have continually tested the resilience and resourcefulness of Mexican society. Droughts have regularly destabilized agricultural production and have affected food security and social and economic well-being in various parts of the country throughout history and prehistory (Florescano, 1980; Hodell et al., 1995; Conde et al., 1997; Liverman, 1999) and remain a problem today. Successive droughts in the 1930s, 1950s and 1990s, for example, contributed to water scarcity, harvest failure, illness, livestock disease, land abandonment and water conflict in the north of the country (Sandoval, 2003). Catastrophic flood events, some associated with hurricanes, have also killed thousands of people across southern and central Mexico, have left tens of thousands homeless and caused billions of dollars of damage, both directly and through rainfall induced land sliding.[1] In short, over three hundred years after the devastating events of the 1690s, the impacts of climate change still rank among the most significant threats to social and economic well-being and environmental security in Mexico.

Current climate models all indicate that 'it is likely that all land regions will warm in the 21st century' (IPCC, 2007: 850). It is thought that significant potential increases in temperature will be accompanied by changes in precipitation, with the possibility of more frequent and intense extreme weather events (Fraser et al., 2003; Bogardi, 2004; IPCC, 2007). While it should be acknowledged that such changes could bring opportunities for some, they will increase the vulnerability of others, especially those already socially, environmentally or politically marginalized sectors of society (Tompkins and Adger, 2004). In Mexico, where a substantial proportion of the population works in an agricultural system that relies on relatively low and variable rainfall, and whose prosperity is critical to the nation's debt burden economy, such conditions could prove particularly disastrous (Liverman, 1999). It is anticipated that most of central America in fact will warm over the next century in line with global mean warming, but annual precipitation is likely to decrease (IPCC, 2007: 892). There are fears that higher temperatures and reduced precipitation could increase the competition for water and falling ground water levels which could in turn exacerbate existing tensions over this most vital resource.

Considerable insight into the specific regional social and economic implications of predicted climate changes can be gained by analysing the spatial vulnerability of society, agriculture and other resources to current and predicted variations in temperature and precipitation (Magaña et al., 1997; Liverman, 1999). The way in which human activities and the natural environment have

been affected by and have responded to climate changes and weather perturbations in the past, however, provides another invaluable guide to where the
most critical vulnerabilities to possible future climate changes may lie. The
purpose of this study is to explore the complex interaction between climate and
society in historical perspective across Mexico, by investigating the nature and
scale of the impacts of climatic variability and unusual and extreme weather
events on different agricultural communities in colonial Mexico between 1521
and 1821. Attention focuses on the variety of responses to climate changes
and weather events and how these responses varied according to the changing
social, political and economic circumstances throughout the colonial period.
The degree to which these impacts and responses were a function of differential social and environmental vulnerabilities will be considered.

Changing Vulnerabilities

Vulnerability can be broadly defined as the potential for loss (Cutter, 1996:
529) or 'the degree to which human and environmental systems are likely
to experience harm due to a perturbation or stress' (Luers et al., 2003:
255). Though frequently referred to and widely used in the risk, hazards
and disaster literature (see, for example, Burton et al., 1993; Blaikie et al.,
1994; Kasperson and Kasperson, 2001), vulnerability is becoming an
increasingly important concept in the fields of environmental and climate
change (Cutter, 2006). Vulnerability is not static. The temporal context of
vulnerability is of critical importance. Yet there have been relatively few historical treatments of vulnerability (Cutter, 1996: 533) and less still of how
vulnerability to climate variability may have changed and may be changing
over time (Parry, 2001: 258).

To some extent, this relative neglect is a function of three key problems:
first, the multiplicity of definitions of and approaches to vulnerability and a
general lack of consensus on the meaning of the term (Luers et al., 2003:
255); second, vulnerability is a concept, not an observable or tangible phenomenon per se, rendering it difficult to identify in the historical record;
and third, we still possess only an imperfect knowledge of how the climate
has changed over the historical period.

Adopting Cutter's terminology, there is certainly a 'confused lexicon' of
meanings, interpretations and approaches to understanding vulnerability
(Cutter, 1996: 530), reflecting a suite of epistemological orientations and
contrasting frameworks. Various genres of literature, for example, have
focused on vulnerability from the perspective of biophysical threats and
hazards (Hewitt and Burton, 1971; Gabor and Griffith, 1980; Ambraseys
and Jackson, 1981; Burton et al., 1993; Blaikie et al., 1994). Many 'unapolo-
getically naturalist', physicalist explanations of disasters, for example, place

total blame on 'the violent forces of nature' (Frazier, 1979; Foster, 1980; Blaikie et al., 1994: 11). These approaches position people as being implicitly vulnerable, owing to their presence in particular fragile or precarious environments. Adopting this perspective, those societies considered most vulnerable to drought, for example, would be those living in areas of low rainfall and sandy soils (Liverman, 1999).

Some more subtle biophysical based assessments have also incorporated demographic factors, such that concern focuses on the capacity of a particular environment to support a population. In recent years there has been a move away from attention on the stressors in vulnerability studies to the stressed system and its component parts and its ability to respond (Ribot, 1995; Clark et al., 2000). Social vulnerability, for instance, focuses on the 'susceptibility of social groups or society at large to potential losses' (Cutter, 2006: 72) from hazardous events and disasters. In these approaches, as Macnaghton and Urry (1998) suggest, 'cultural criteria are implicated in the definition and trajectories of even the most apparently physical environmental issues' (cited in Jones, 2002: 248). The nature of a hazardous event is taken as a given or at the very minimum viewed as a social construct, not a biophysical condition (Cutter, 1996). This perspective thus highlights human coping strategies and responses, including societal resilience to environmental hazard, and positions biophysical threats as socially constructed phenomena. The most extreme of such approaches have been criticized for environmental denial (Burningham and Cooper, 1999: 306) and for fostering political quietism with respect to pressing environmental issues (Milton, 1996: 54; Demeritt, 2001).

Interdisciplinary research teams have now begun to explore vulnerability as a function of exposure, sensitivity and adaptive capacity, manifested within interactions of social and ecological systems (Turner et al., 2003). Vulnerability in this body of literature is conceived both as a biophysical risk as well as a social response (Cutter, 1996: 533). The impacts of an extreme climate event for a community, for example, would then be influenced 'as much by the level of technological, economic and political development as by the severity of the meteorological event itself' (Liverman, 1990: 49). The complex interaction between the environment and human society, however, represents a constraint that not only influences how livelihoods in a region may be vulnerable to disruption, and the way in which environmental change or an environmental or climate event is experienced, but also how different social systems and groups can respond or adapt to such events (Oliver-Smith and Hoffman, 1999: 73; Ohlsson, 2000).

Kasperson et al. (1988) and Kasperson and Kasperson (1996), for example, suggest that risks and hazards interact with cultural, social and institutional processes in such as way as to temper or aggravate public response (Cutter, 2006). Vulnerability, and hence adaptability, may thus change according to different scenarios and over time. Trajectory analysis can go some way to revealing these regional dynamics of change and, it follows, changing vulnerabilities (Kasperson et al., 1995). Messerli et al. (2000), for example, have suggested that the vulnerability of a particular society changes over time in conjunction with adaptive mechanisms and adjustments, some of which may in fact render society more vulnerable in the long run. They argue that there is a historical 'trajectory of vulnerability' through which all societies pass as they develop, economically, socially and technologically. The position of a society on this trajectory influences its relative vulnerability. 'Nature dominated' hunting and gathering societies, for example, are considered to be the most vulnerable to environmental and climatic changes. Societies who have modified their environments in order to buffer themselves from these changes, however, and who have developed productive agrarian-urban systems based on these adaptations are thought to be more resilient. With population expansion, however, and the overexploitation of and reliance upon these buffer systems, it is thought that the vulnerability of these societies once again increases. There has been considerable debate as to whether modern society is effectively becoming more or less vulnerable as a result of technological innovation and adaptation over time (see Meyer et al., 1998: 240–241; Messerli et al., 2003). Only through regionally focused, historical analysis of vulnerability, however, can such questions be adequately addressed.

Climate Change and the 'Double-Sided' Structure of Vulnerability

There is now little doubt that the global climate is changing and human activities are exacerbating natural climatic variability. That the climate to which the world is currently accustomed will undergo further significant change is also indisputable. The details and implications of these changes, however, are less clearly understood. Such changes in climate are likely to pose new and significant challenges for society at large (Adger, 2003: 387) and for this reason climate change and its implications are increasingly being recognized as the most important influences on both biophysical and social vulnerability (O'Brien and Leichenko, 2000).

Unusually severe or prolonged drought, for example, ranks among the most devastating and calamitous of all extreme climate events, contributing to wildfires, crop failure, livestock death, food shortages and famine. Yet the vulnerability to extreme droughts is also thought to be increasing in many parts of the world (Baethgen, 1997), judging by escalating economic losses and a greater number of fatalities due to such events (Meyer et al., 1998; Kundzewicz and Kaczmarek, 2000; Easterling et al., 2001). Understanding why vulnerabilities are changing and how and why certain sectors of the population are disproportionately affected by drought, and investigating how people respond to its effects, are considered critical steps toward designing appropriate drought preparedness, mitigation and relief policies and programmes (ISDR, 2002).

Floods are also considered to be among the most common and yet most damaging of all extreme events. No other natural hazard occurs as frequently around the world or causes as much collateral damage (Kunkel et al., 1999; Kundzewicz and Kaczmarek, 2000; Changnon et al., 2000; Berz et al., 2001: 458; Kohle and Dandekar, 2004). The frequency and magnitude of floods from unexpected and unusual rainfall, but also associated with hurricanes and windstorms, is expected to increase in the next few decades in a context of predicted global warming and sea level rises (Kundzewicz and Kaczmarek, 2000; Changnon et al., 2000; Yin and Li, 2001; IPCC, 2007: 6). Yet vulnerability to floods and, it follows, flood losses, is also a product of anthropogenic change. Increasing deforestation, urbanization and river regulation and decisions to settle floodplains either through choice or due to land pressure, have rendered people in some locations effectively more vulnerable to flood risk. Moreover, although changes in technological development and agricultural practices can help to create a society that is effectively better able to survive the impacts of such events, they can also lead to over-dependence upon these systems. Indeed, it has been argued that a reliance on existing flood alleviation schemes has actually encouraged people to settle, build on or exploit lands in even higher risk zones (Kundzewicz and Kaczmarek, 2000). As suggested above, combined with population expansion, such developments can and will effectively increase the level of social vulnerability to the impacts of flooding over longer timescales (Messerli et al., 2000).

Climate change and weather events could also have important health implications. Climate can constrain the range of infectious diseases and can also influence pathogens, vectors, host defences and habitat, and many diseases are thought to be influenced by weather conditions or show strong seasonality (Patz et al., 2000, 2005). Global warming, however, is thought to be playing an increasingly important role in driving the global emergence and redistribution of infectious diseases, while extreme weather events may influence the timing and intensity of infectious disease outbreaks (Epstein, 1999, 2001, 2002; Patz et al., 2005). Variation in the incidence of vector borne diseases, for example, has been strongly associated with extreme weather events and annual changes in weather conditions (Zell, 2004). El Niño Southern Oscillation (ENSO) events in particular are often accompanied by weather anomalies that have been strongly associated with disease outbreaks over time (Epstein et al., 1997; Epstein, 1999; Pascual et al., 2000; Kovats et al., 2003). There is considerable debate about the precise nature of the health impacts of climatic changes, the nature and timing of the effects and their possible beneficial or adverse consequences (McMichael et al., 2006). Identifying the causal associations between climate, extreme events and disease outbreaks remains a major challenge (Kovats et al., 2003: 1482) and a better comprehension of the linkages between climate variability and disease should be seen as imperative if predictive models to guide public health responses are to be devised (Parmenter et al., 1999: 814).

Climatic changes will thus disrupt the way people live and interact with their environment and with each other (Chen and Kates, 1994: 3) and will pose challenges to future livelihood strategies for everyone (Bohle et al., 1994). Of course, unpredicted changes or extreme weather events, 'when climate surprises occur in unexpected places or with unexpected frequency' (Streets and Glantz, 2000: 99), have the potential to affect, and indeed disrupt, even the most robust of environmental, social and economic systems. It is anticipated, however, that the impacts of future climate changes are most likely to be felt more severely by natural resource dependent communities (Adger, 2003), and particularly by those societies in the developing world (Liverman, 1999). Agricultural economies are indeed thought to be more vulnerable to climatic hazards than industrialized ones, while conditions are most critical for environmentally, politically, socially or economically marginalized communities (Meyer et al., 1998; Cutter, 1996; Liverman, 1999), whose ability to adapt to and recover from environmental changes and biophysical events can be limited or constrained by limited access to natural resource or financial capital (Fraser et al., 2003; Mirza, 2003). People who live in arid or semi-arid lands, in low lying coastal zones, in water limited or flood prone areas, or on small islands are obviously especially vulnerable to climatic variability and any changes therein. The very young and very old are also often regarded as especially vulnerable, while differences in health, or access to health facilities, ethnicity, education, and learned experience with the hazard in question can all influence vulnerability (Meyer et al., 1998). Moreover, women in some parts of the world might also be differentially more vulnerable than men to environmental changes, particularly in developing countries where they are often responsible for agriculture, fuel wood and water management and hence are disproportionately affected by drought, deforestation and water pollution (Liverman, 1999).

Inasmuch as many climate change predictions have tended to be at best pessimistic and at worst apocalyptic (Liverman, 1999: 112), the predicted impacts for all these different sectors of society are similarly gloomy. Past climate impacts research, for example, and much of the continuing policy discussion has viewed climate change as something of a Malthusian threat to the world's overall ability to produce enough food and support a rapidly expanding population (Chen and Kates, 1994: 4). Vulnerability to climate change has been considered to be the very 'key to human security' (Bogardi,

2004: 362). Indeed, climate change, and an increasingly 'weaponized' world, have been identified as the two most important challenges facing the world in terms of international security (McBean, 2004: 183). Environmental security discourses, for instance, have drawn links between climate change, environmental degradation, economic decline and the debt crisis, and have positioned climate change as a high priority security concern, with the potential to stimulate or aggravate conflict and tension both within and between nations (Homer-Dixon, 1991, 1995). Moreover, by extension, the implications of climate change have also been highlighted as a potential cause of environmental migration, contributing to the movement of thousands of 'environmental refugees' across borders (Barnett, 2003).[2]

Although climate change undoubtedly poses risks to human welfare and as such represents a security concern, literature on such themes has been criticized for being more theoretically than empirically driven and for being motivated by predominantly northern geostrategic interests (Barnett, 2003: 8). Furthermore, the overemphasis on pessimistic prediction has tended to overshadow any net benefits that might be derived from climatic changes measurable through increased agricultural productivity, improved availability of key resources such as water, a reduction in some risks, such as those associated with flooding or decreased climate related expenditures. There has been a general reluctance among both scientists and policy makers to discuss the existence of any such positive outcomes lest they undermine efforts to secure global consensus on the implications of climate change (O'Brien and Leichenko, 2000: 223; 2003) and rightly so. This general pessimistic stance, however, has also, to some extent, obscured the fact that vulnerability to climate change does have an orienting function and that human societies are adaptable and have developed institutions and cultural coping strategies to deal with the impacts of environmental changes (Hassan, 2000; Fraser et al., 2003). This is particularly true of extreme or 'surprise' events which can, under certain circumstances, provide a 'window for positive change and impetus for remedial action' (Streets and Glantz, 2000: 100).

On this theme, Vogel (2001) has suggested that greater emphasis needs to be placed on the 'double sided structure' of vulnerability, that is to say, studies of vulnerability that also pay attention to adaptive capacity. Adaptation in this sense can be defined 'as an adjustment in ecological, social or economic systems in response to observed or expected changes in climatic stimuli and their effects or impacts in order to alleviate adverse impacts of change or take advantage of new opportunities' (Adger et al., 2005). While most adaptation is reactive, in response to past or current events, it can also be anticipatory and planned (Smithers and Smit, 1997; Smit et al., 2000; Pelling and High, 2005: 1). Moreover, societies have an inherent capacity to adapt to climate change and these capacities are often bound up in an ability to act collectively (Adger, 2003: 388). Recent work focusing on 'social capital' responses has highlighted that environmental problems may be as likely to promote community co-operation as much as conflict between communities (Liverman, 1999: 112). Community engagement may thus offer a means of reducing vulnerability to natural hazards associated with climate change (Abramovitz et al., 2001). Although there has been an 'explosion of interest' in such studies of adaptation to climate change in recent years (Adger et al., 2005), there is still a need for a more comprehensive understanding of the pathways through which social resources are accumulated, on the influence of political structure on decision-making processes within communities and on the role of institutions and culture in shaping adaptive capacity and action (Pelling and High, 2005: 2).

Vulnerability is thus a composite and dynamic concept which means different things to different people (Cutter, 2006). It can be discussed in ecological terms, in relation to political economy or class structure and as a reflection of social relations and, in as much, represents a multi-layered and multidimensional social space defined by the determinate political, economic and institutional capabilities of particular groups of people in specific places at specific points in time. It follows that assessments of vulnerability need to be capable of mapping the historically, spatially and socially specific realms of choice and constraint that determine risk exposure, coping and adaptive capacity and recovery potential (Bohle et al., 1994). Such complex mosaics inevitably cast doubt on attempts to describe any general patterns and trends at global scales (Liverman, 1990) and justify the need for regionally or locally focused studies which are necessarily likely to be much more revealing. Yet, to date, relatively few studies have adopted such a multi-factor regionally based approach and fewer still compare the changing vulnerability of one place to another (Cutter, 2006: 116).

There are, therefore, still considerable gaps in our knowledge of climate changes themselves, but also exposure, sensitivity, adaptability and vulnerability of social and environmental systems to any possible changes in the future (Parry, 2001: 258). We have only an incomplete understanding of differential social vulnerability in different places through time and are only partially aware of the ways in which adaptations at different scales of human organization may interact and in some cases lead to differential or unexpected effects. Moreover, we do not know the extent to which adaptation might help overcome or exacerbate the diverse threats climate change poses in a more crowded and demanding world. Further work is needed to identify and integrate information about exposures, sensitivity and adaptability in the past in order to provide more detailed information about the potential impacts and responses associated with future climate changes (Parry, 2001: 258).

Exploring Climate and Society in Mexico

Mexico represents one of the most climatically sensitive areas of the world (Wallen, 1955; Kutzbach and Street-Perrott, 1985; Liverman, 1993). The country lies in a latitudinal belt that is sensitive to fluctuations in atmospheric circulation and has, accordingly, experienced climatic change on the long (Heine, 1988; Bradbury, 1989; Metcalfe et al., 1991) and short timescales (García, 1974; Jáuregui and Kraus, 1976; Jáuregui, 1979; Metcalfe, 1987; O'Hara and Metcalfe, 1995). Such changes can be explained by present day climatic mechanisms.

The Mexican climate is influenced by two dominant atmospheric circulation systems. These are the trade winds, which deliver summer rainfall to the central and southern regions of the country when the Intertropical Convergence Zone (ITCZ) shifts northwards, and the subtropical high pressure belt, which brings stable dry conditions to the country during the northern hemisphere winter, when the ITCZ is displaced equator-wards. The northern part of the country is climatically distinct from the central and southern regions and tends to be marginal to the summer trade winds, though the northwest of the country is affected by the westerlies during the boreal winter, which can bring rainfall to this region.

The distribution and total annual rainfall across the country are determined by shifts in the strength and location of these dominant atmospheric circulation systems. Seasonality of precipitation, and variations therein, and the incidence of tropical storms also dictate changes in the temporal and spatial distribution of precipitation. Some drought events may relate to summer high pressure which can disrupt the flow of moist air, creating a mid wet season drought or canicula (Liverman, 1999).

A number of other climatic features play an influential role in the distribution of rainfall and annual precipitation totals. These include hurricanes and tropical cyclones, which typically affect both Caribbean and Pacific coasts between May and November, with September having the greatest frequency. Mid-latitude cyclones can bring rain, hail, sleet and snow. Severe thunderstorms, which form along or ahead of cold fronts, are also common during the spring and summer months and are often accompanied by hailstorms (Mosiño Alemán and García, 1974). ENSO events are thought to represent the most significant cause of inter-annual climate variations across the country. Evidence of ENSO related rainfall anomalies for the instrumental period, for example, suggests that during warm ENSO (El Niño) events, an enhancement of the mid-latitude westerlies brings above normal rainfall to northern Mexico in the normally dry season between November and April, while a reduced flow of air across the country with the ITCZ well to the south results in drought and heavy frosts in central Mexico (Ropelewski and Halpert, 1987, 1989; Hastenrath, 1988; Cavasoz and Hastenrath, 1990; Magaña, 1999; Magaña et al., 2003). Cold (La Niña) events have been linked to lower than average winter rainfall in northern Mexico and higher than average summer rainfall in central Mexico. Climatic anomalies associated with recent ENSO years have thus resulted in widespread harvest failure and livestock losses, while there is also thought to be a link between hurricane frequencies and ENSO years.

Topographic features of course also exert an important influence on climatic characteristics and variation across the country. The Mexican plateau or high plain (Altiplano), which has an average elevation of 1,500 metres above sea level, is flanked on the western edge by the Sierra Madre Occidental and on the east by the Sierra Madre Oriental. These cordilleras can trap stable air masses, preventing them from entering the central Mexican plateau, and can also deflect winds. Trade winds from the northeast, over the Gulf of Mexico, for example, can be deflected southwards (through the Tehuantepec Pass), where they can become quite violent northerly blowing winds. Topography also plays a critical role in thermal variation and hence also rainfall distribution. Adiabatic cooling of air as it ascends over the cordilleras leads to the release of moisture in the form of heavy rains over the mountain slopes, this particularly being the case with the Sierra Madre Occidental, which is oriented at right angles to the surface trades that blow over the Gulf of Mexico. The extensive Altiplano also exerts an important thermal effect, heating air masses, leading to convective activity and thunderstorms during the summer rainy season (Mosiño Alemán and García, 1974). Frosts are also relatively common at high elevations, particularly between October and April. Dry atmospheric conditions in the Altiplano, however, can result in radiative cooling and frosts even during the summer months (Morales and Magaña, 1999). Indeed, frosts in June or July, such as those recently recorded in central Mexico (Eakin, 2005: 1926), can threaten rainfed crop yields.

Climate History and Vulnerability in Mexico

Climate changes on a range of timescales have had implications for social and economic well-being throughout Mexican history and prehistory. Yet by the time Hernando Cortes and the first Spanish conquistadores arrived in Mexico in 1519, the region had been the site of some of the most advanced civilizations of the western hemisphere. A number of scholars have investigated the nature and diversity of landscape changes associated with the climate risk avoidance strategies of such 'traditional' and pre-Hispanic agrarian societies (Wilken, 1987; Whitmore, 1992). The most significant agricultural adaptations to climate variability, periodic drought and the constant threat of water scarcity for agrarian communities, however, involved the storage of water and the use of irrigation. Permanent and ephemeral watercourses, rivers and arroyos had long been exploited by pre-Hispanic populations for irrigating *milpas* (maize plots), garlic and beans.

Though for the most part successful, such adaptations may have failed to buffer societies against the most extreme weather. Prolonged drought coupled with 'killing' frosts in central Mexico between 1452 and 1455, for example, is thought to have contributed to widespread harvest loss, culminating with the famine of '1 Rabbit'. This stimulated out-migration, disease and death (Hassig, 1981; Therrell et al., 2004). Though undoubtedly one of the most widely reported social calamities in Mesoamerican history, this event was not unique (Therrell et al., 2004) and similarly dramatic episodes have been documented both before and after this time.

Following the conquest of the Aztec capital at Tenochtitlán in 1521, Spain was to rule Mexico as part of the viceroyalty of New Spain for the next 300 years until 16 September 1810, when the battle for independence (finally gained in 1821) began. Although many of the strategies developed to cope with the threats posed by the Mexican climate were maintained in the colonial period, and other adaptations and coping strategies would be developed, the features of the colonial political economy are thought to have created a society and an economy that was differentially more vulnerable to the impacts of climate change and weather events (Liverman, 1999).

The timing, impacts and responses engendered by historical climate variations and weather related events in Mexico have only recently begun to be investigated. A key obstacle has been the lack of adequate, long-term data with sufficient resolution to investigate changes in local weather conditions and also the social and economic implications of these changes (Ingram et al., 1981; Easterling et al., 2001; Houghton et al., 2001). Short-term extreme events in particular require high temporal resolution observations, but widespread instrumental weather data rarely cover periods before the middle of the nineteenth century (Metcalfe, 1987).

Investigations of lake basins in the north of Mexico and in the neovolcanic highlands of west central Mexico, however, have revealed significant spatial variability in climatic signals across Mexico over the late Pleistocene and Holocene, resulting from the influence of different moisture sources in response to external forcings. Northern and western Mexico came under the influence of Pacific mid-latitude frontal systems during the last glacial. The Holocene also shows high variability, with changes (especially drought periods) that may have contributed to shifts in population distribution (Metcalfe et al., 1994).

Palaeolimnological analysis of sediment cores taken from lakes elsewhere in Mexico has been similarly revealing. Analysis of lake sediments from

Punta Laguna in the Yucatan Peninsula, for example (and also from Lake Peten Itza in the lowlands of Guatemala), has indicated climatic changes on multi-decadal and millennial scales (Curtis et al., 1996, 1998; Hodell et al., 2005a). The period between 1785 and 930 years BP was thought to be particularly dry (with exceptionally arid events centred around 862, 986 and 1051 AD) and is recorded at several other sites in Mexico and Central America more generally. This period coincides with, and indeed has been linked by some authors to, the collapse of the Classic Maya Civilization, when urban centres faced depopulation and trading decreased across a broad region (Curtis et al., 1996).

Other, more recent periods of climatic change have also been recognized from similar sources. Analysis of sediments drawn from Lake Aguada X'caamal in the northwest Yucatan Peninsula has revealed that the climate became drier in the region in the fifteenth century AD, round about the time of the onset of the Little Ice Age. Comparison of the results from the Yucatan Peninsula with other circum-Caribbean palaeoclimate records indicates a coherent climate response for this region at this time (Hodell et al., 2005b).

The expanding network of tree ring studies is also playing a pivotal role in the reconstruction of Mexico's climate history. Tree ring investigations by Cleaveland, Stahle, Therrell and Villanueva Diaz have provided superb records of historical precipitation variations in various regions across the country. Dendroclimatological evidence supports lake sediment evidence of prolonged periods of drought, or megadrought, in the eighth and ninth centuries (Acuña-Soto et al., 2005), while Cleaveland et al. (2003), Dias et al. (2002) and Therrell et al. (2006) have highlighted other periods of sustained, extreme drought across northern Mexico which are thought to have affected agrarian livelihoods and may have also contributed to social and political instability. Megadrought in the 1550s, for example, which is thought to be one of the most severe droughts in North American history, has been forwarded as an underlying cause of indigenous revolt at this time (Cleaveland et al., 2003: 370). This megadrought may have also interacted with prevalent ecological and sociological conditions, magnifying the human impact of infectious disease in central Mexico (Acuña-Soto et al., 2002). Interestingly, the occurrence and impacts of such extreme droughts in central Mexico has been found to corroborate Aztec climate folklore (Therrell et al., 2004).

Historical documents represent another source of high quality climate information at the fine temporal resolution needed to identify past climate fluctuations (Brown and Isaar, 1999; Bradley, 1999). Historical records charting the impacts of extreme events, coping strategies, technological adaptation, narrative, ideology, regulation and recovery have long been recognized as valuable sources for reconstructing climate during the last few centuries when no instrumental or similar sources are available

(Garcia-Herrera et al., 2005). Indeed, over the past two decades, climate historians working in various regions of the world have refined and improved methodological approaches and have identified new documentary and proxy sources of climate information. Such sources have been used extensively to reconstruct historical ENSO events (Quinn and Neal, 1992; Ortlieb, 1999), as well as changes in the North Atlantic Oscillation (Jones et al., 1997; Luterbacher et al., 1999) and there have been a number of documentary based studies of regional climate variability across Europe,[3] the Americas,[4] Africa,[5] and Asia[6] and also over the Pacific and Atlantic Oceans.[7]

Inasmuch as historical documents can be used for the reconstruction of climate variability, however, they can also offer invaluable and, to some extent, unique insight into how people perceived, were affected by and reacted to a variety of climate changes and associated impacts. Travel accounts and descriptions, legal documents, crop and tax records, as well as maps, paintings and images have all been used to identify the timing and chart the impacts of and societal responses to anomalous weather and extreme events over the historical period in various parts of the world. The very rich colonial archives of Mexico present similar opportunities. The Spanish colonial administration, both in Spain and in Mexico, began compiling a series of document collections on population, settlements, landscapes and economic resources for the different regions of Mexico immediately after contact (Butzer, 1990). This has resulted in a legacy of archival data in the form of national and regional level fiscal and judicial documents now housed in the various national, state level and local archives and libraries of the country.[8]

A number of regional archival investigations have highlighted the different document groups (ramos) and historical sources that can be employed to reconstruct microscale environmental characteristics and to identify the environmental impacts of post-Conquest changes in land use and tenure (Melville, 1990, 1994; Butzer and Butzer, 1993, 1995, 1997; Endfield and O'Hara, 1999; Sluyter, 2001, 2002). Investigations have also used these sources to explore the relationship between agricultural and economic crises and periods of drought (Swan, 1981; Florescano et al., 1995), and to investigate the connection between water shortage and conflict over water access and water rights (Lipsett Rivera, 1990, 1992, 1999a, 1999b; Meyer, 1997; Endfield and O'Hara, 1997). Recent research has also demonstrated the potential of these colonial documents for reconstructing climatic chronologies and for investigating the impacts and responses engendered by extreme weather events in different regions of the country over the last six centuries (Metcalfe, 1987; O'Hara, 1993; García-Acosta, 1993; O'Hara and Metcalfe, 1995; Endfield et al., 2004a, 2004b; Endfield and Fernández-Tejedo, 2006). The present study makes use of a variety of colonial archival sources, housed in national and regional, public and private archives of Mexico, to further explore the complexities of the relationship between climate and society in colonial Mexico.

Case Studies and Approach

This study focuses on three regions covering a range of environmental, social, economic and political contexts and histories and located at key points along a north-south rainfall gradient: the Conchos basin of Chihuahua in the arid north, the central Valleys of Oaxaca in the wetter south and Guanajuato located in the central highlands, a region of climatic transition. As shall be illustrated in the following chapters, each region had very different environmental and climatic contexts and pre-Hispanic histories and developed very different settlement and land use characteristics during the colonial period. Guanajuato had a long history of pre-Hispanic settlement, but by the time Spanish arrived, the area represented something of a frontier zone. Nonetheless, the combination of fertile soils and mineral wealth meant that by the second half of the sixteenth century, the area had developed into the country's main agricultural centre, specializing in the production of wheat, fruits and vegetables. In contrast, the central valleys of Oaxaca retained much of their pre-Hispanic indigenous character throughout the colonial period and, although livestock and Mediterranean cultivars were introduced into both areas, people there continued to focus on the production of maize, beans and chile, employing traditional cultivation methods. Significantly, there was a good deal of land retention by indigenous populations in this region. In Chihuahua, the threat of violent attacks by the hostile nomadic indigenous groups who occupied the northern part of Mexico delayed European colonization of the region. By the start of the seventeenth century, however, a lucrative mining and livestock economy had begun to develop. Investigating the relationship of climate and society through these case studies might shed light on the complex climatic relationship between northern, central and southern regions of the country, but also provides a unique insight into the way in which vulnerabilities and adaptations to disaster changed through time and across different contextual spaces. The case studies selected also provide an opportunity to explore how predominantly indigenous societies, as well as those with a more European structure, understood, coped with and articulated knowledge about climate change.

The nature and origin of the documents that can be used to investigate climate variability, impacts and implications for society in these regions are extremely diverse and require some consideration. The Archivo General de la Nación (AGN) is home to one of the richest manuscript collections on the history of the Americas. There are in total 115 ramos (consisting of over

300,000 individual documents) dealing with colonial New Spain alone now comprehensively indexed on a CD-ROM index. A variety of ramos held in the AGN and noted in table 1.1 were consulted for this study. As well as the manuscript collections held in the national capital, there is also a rich store of state level and local archives, both civil and ecclesiastical. The present study made use of unpublished, handwritten documents held in several private and public archives in the case study regions, including the Archivo General del Estado de Oaxaca (AGEO), Archivo Historico Municipal de Oaxaca (AHMCO) and the Archivo Privado de Don Luis de Castañeda, all in Oaxaca City; the Archivo Historico Municipal de Leon (AHML) and the Archivo Historico del Estado de Guanajuato (AHEG) in Guanajuato; and the Archivo Historico Municipal de Fondo Colonial, Chihuahua (AHMCH) and Archivo Hidalgo de Parral, Chihuahua (AHP). A number of the many libraries and museums in Mexico City also have archival repositories open to public consultation. A variety of published texts and unpublished documents now filed on microfilm were consulted in the library of the Museo Nacional de Antropología, including original documents, and photocopies of documents, dealing with all three case studies, that are now housed in libraries in Paris (Bibliothèque Nationale de Paris) and Spain (Archivo General de Indias). Tables 1.1 and 1.2 demonstrate the range of ramos (document groups) that were employed in this investigation and the nature of the typical content.[9]

Manuscripts in each ramo were read in chronological order. Documents were scrutinized for references to weather or weather related information, falling into three approximate classifications: first, direct descriptions of extreme or damaging weather events such as drought, floods, storms, 'killing' frosts, hail storms; second, general observations of contemporary climatic conditions, such as the early, late or non-arrival of the summer rain season, seasonal frosts, changes in the characteristics of perceived 'normal' weather conditions and the implications of these, particularly for agricultural communities, but also for health and social well-being more generally; and third, indirect evidence of climate variability, such as that derived from information on harvest gains or losses, problems of food or water scarcity, disputes and legal proceedings over natural resources (including water), instances of crop blight, marked variations in grain prices, migration, land or property abandonment, pestilence and epidemics (human and livestock). Relevant passages were all excerpted and transcribed. A wide variety of published chronicles, diaries, journals and travelogues were also consulted for references and observations on weather and its implications.[10] Wherever possible, every effort was made to corroborate evidence with multiple references, and to compare archival with other lines of climate change evidence derived from other written sources or, where available, independent dendrochronological and archaeological evidence of climate variability and change.[11]

Table 1.1   Document groups (ramos) consulted in the Archivo General de la Nación (AGN), Mexico City

Ramo	*Content*
Tierras	Disputes over land and water/natural resources
Mercedes	Land grants
Indios	Indigenous affairs
Ayuntamientos	Local authority records
Rios y Acequias	Information on rivers and water works
Caminos y Calzadas	Information on roads and routeways
Historia	Historical information for various regions, compiled at close of eighteenth century
Obras Publicas	Documents pertaining to public works projects
Industria y Comercio	Materials pertaining to trade and commerce of various goods/trading infrastructures and networks
Tributos	Indigenous tax records
Alhóndigas	Crop records/information on goods held in public food store
Audiencia de Mexico	Documentations relating to the administration of New Spain
Archivo Historico de Hacienda	Affairs of haciendas and landholdings/includes correspondence
Hospital de Jesus	Disputes judicial and administrative documents. Materials pertaining to the administration of the Marquesado del Valle, Oaxaca (see chapter 3) were consulted
Jesuitas	Information on Jesuit property
Temporalidades	Information on Jesuit property subject to redistribution after their expulsion in 1767
Carceles y Presidios	Prison and garrison records
Provincias Internas	Documents pertaining to military command of northern frontier
Misiones	Mission records
Alcades Mayores	Provincial level administrative records
Media Anata	Tax paid tied to incomes derived from any ecclesiastical benefit, pensions or use
General de Parte	Requests, complaints and demands forwarded to Viceroy and/or local council

There are of course a number of problems with using archival sources for investigating the impacts and perceptions of and vulnerability and response to extreme events. In the absence of independent climatic evidence, recorded information of climate variability and weather events cannot at once serve

Table 1.2 Document groups consulted in the regional archives in Oaxaca, Guanajuato and Chihuahua

Regional archive	Ramo	Content
Archivo General del Estado de Oaxaca (AGEO)	Alcadías Mayores and Real Intendencias	Local, political and economic issues, litigation over resources and land, correspondence between hacienda owners, property inventories, details of harvest problems
Archivo Historico Municipal de Oaxaca (AHMCO)	Actas de Sessiones de Cabildo	Municipal council minutes
Archivo Privado de Don Luis de Castañeda	Not named	Various. Focused on hacienda documents, deeds, transactions, maps and legal disputes over land
Archivo de las Notarias, Morelia, Michoacan (ANM)	Tierras y Aguas	Disputes over land and water
Archivo de la Ciudad de Patzcuaro (P)	Not named	Various documents pertaining to barrios and pueblos in the region
Archivo Historico Municipal de Léon, Guanajuato (AHML)	Notarias	Various documents pertaining to periods of agrarian crisis and impacts
Archivo Historico del Estado de Guanajuato (AHEG)	Actas de Cabildo Ayuntamiento	Municipal council minutes Local authority record
	Actas de Cabildo Guerra	Municipal council records Details of defence needs and indigenous attacks on northern frontier of Mexico
	Provincias Internas	Information on unrest on the northern frontier of Mexico
	Haciendas	Information on properties and landholdings in Chihuahua, plus information on epidemic and agrarian crisis

as both meteorological evidence and evidence of the human impacts of climate (De Vries, 1980). The purpose in this study is not to produce a chronology of climate changes or events per se, but there are other problems associated with subjectivity, bias, discontinuity in the historical record and inconsistency that need to be considered. As primary sources written by individuals from particular perspectives and for a specific intended audience, the colonial records used in this investigation, for example, were produced within a certain context and contain both intentional and unintentional biases. There may well have been instances where there was deliberate sensationalism of the impacts of particular flood or drought events in order to secure financial aid, tax relief or to challenge existing rights of access to water. There is also no uniform and agreed definition of what constituted a weather or weather-related event such as a flood, storm, hurricane or drought. Unusual, anomalous or extreme events were judged against what was perceived to be a normal range of variations, which itself was a function of the nature and span of an individual's experience and the average range of variation communicated through oral histories or historical knowledge (Hassan, 2000).

Risks interact with social, cultural, economic, psychological and institutional processes in ways that either amplify or dampen public responses (Kasperson et al., 1988). Society selects which risks or hazards to emphasize and which ones to ignore, often reflecting moral, political and economic choices that are in themselves socially constructed and value laden. Thus, if the climate variability or an unexpected weather event resulted in only limited human impact and economic loss, then there may well be no record of it having taken place (Douglas and Wildavsky, 1982). There may, in addition, have been periods when events themselves caused sufficient disruption in the administrative systems responsible for record keeping to lead to a gap in the record for a period when data is most needed (Landsberg, 1980). For all these reasons, archival investigations of the implications of climatic variability and extreme weather events will inevitably be subject to error.

The colonial archives thus provide at best only a partial record of past events. Furthermore, deriving insights about the contemporary and future vulnerability of Mexican society from such dramatic examples is, of course, problematic. Problems of chronological resolution, insufficient data and oversimplification can hamper the value of using such historical examples as analogues to glean insights about the potential implications of contemporary or predicted climate change scenarios (Meyer et al., 1998; McNeill, 2005: 178). The fact that past societies differ markedly from those in the modern world makes simple analogies or parallels unrealistic (Ingram et al., 1981: 5; Meyer et al., 1998). It should also be remembered that major events, particularly those which leave discontinuities in the historical record, do not always require major causes (Coombes and Barber, 2005: 303) so much as a suite of social, economic, political, demographic and environmental factors that have the potential to coalesce at a particular point in time to cause dislocation, and invariably only then for a particular culture group or sector of society.

Yet equally, while there is a good deal of uncertainty with respect to climatic futures, the value of looking to the historical record to learn about societal vulnerability to climate events should not be underestimated (Bradley, 1999), especially in climatically sensitive regions of the world such as Mexico. Exploring the experiences of individuals, groups and places in the past could help us understand how flexible (or rigid) societies are, or have been, in dealing with climate related environmental changes and might also help us to identify the most vulnerable societies and places (Meyer et al., 1998: 218; Swetnam et al., 1999). Indeed, knowledge of successes and failures in adaptation to past climatic variability might possibly increase the ability to respond to the threats of long-term climate changes (Tompkins and Adger, 2004). Thus, inasmuch as there is a need to develop more comprehensive assessments of changes in weather and extreme climate events over longer timescales (Jones, 1988: 544; Vogel, 2001), it is also vitally important that the impacts of and social response to these changes are explored. It is hoped that the following chapters reveal that, as expressions of contemporary environmental awareness and perception, such documentary sources can offer some insight into how different sectors of colonial Mexican society conceptualized, understood, adapted and responded to climate change and its implications.

Chapter Two

Climate, Culture and Conquest: North, South and Central Mexico in the Pre-European and Contact Period

Environmental Marginality and Society in the Conchos Basin, Chihuahua

Location and climatic characteristics

The modern-day state of Chihuahua (figure 2.1) lies in the northwest central plain of Mexico and is bordered to the north and northeast by the United States, to the west and south by the state of Sonora, and to the southeast by the state of Sinaloa. Covering a surface area of 244,938 sq km, it is the largest state in Mexico. The area can be divided into two physiographic regions, the Sierra Madre Occidental and the basins and hills of the Chihuahua Desert, though each region is characterized by great diversity and can be further subdivided into a number of provinces based on geomorphological and physiographical parameters (Schmidt, 1992).

The Chihuahuan climatic area is part of an arid zone which includes the southwest USA, but the climate of the region per se actually ranges from arid and semi-arid to tropical and subtropical humid conditions (Schmidt, 1992). Average annual temperatures vary across the region from 13° C to 24° C, though summer temperatures in some parts of the state can regularly reach over 30° C and there are occasional extremes of temperature (INEGI, 2003). Rainfall is predominantly seasonal, with two thirds of the precipitation falling during the summer months between July and October, as a true monsoon system. Storms coming from the west, however, can result in winter precipitation in the north and the south (Dettinger et al., 1998; Higgins et al., 1999; Cleaveland et al., 2003). Average annual rainfall is around 350–400 mm per annum during normal years (INEGI, 2003), though rainfall averages vary markedly across the region and from year to year. Indeed,

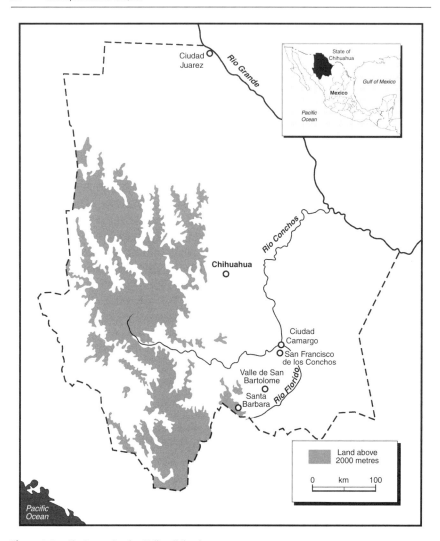

Figure 2.1   The Lower Conchos Valley, Chihuahua

Chihuahua is one of the most seriously and frequently drought affected regions of Mexico (Garcia, 2000). Some of these droughts are thought to be strongly influenced by ENSO (Cavazos and Hastenrath, 1990; Allan et al., 1996; Stahle et al., 2000; Magaña, 1999; Liverman, 1999).[1]

Failure of the summer rains can have, and has had in the past, dramatic and often disastrous implications. Dry soil and high surface temperatures increase evapo-transpiration, which can have devastating implications for agricultural production and the availability of pasturage for grazing.

Successive droughts have contributed to harvest failure, livestock disease and water conflict and have served to highlight the particular vulnerability of society in this region to climatic variability (Sandoval, 2003). During the prolonged drought period between 1948 and 1964, for instance, which is thought to be one of the worst droughts in the area in recorded history (Dias et al., 2002), trees, grasses and cattle died, and ranches were abandoned. Similarly, droughts in 1987 and 1988 greatly reduced crop yields but also forced the sale of livestock, contributed to temporary migration of some sectors of society seeking work as wage labour, and also led to increased usage of non-cultivated plant resources for food and grazing purposes (Laferrier, 1992). Rainfall in the 1990s was also significantly below normal, resulting in irrigation cutbacks and serious agricultural losses (Kim et al., 2002). This drove small-scale farmers off the land and once again severely affected the livestock industry. Recent investigations of the frequency and spatial characteristics of drought in the Conchos River Basin, in southeast Chihuahua, suggest that such prolonged droughts may have a return period of about 80–100 years (Kim et al., 2002). It is thus vitally important to understand how populations have been affected by such events in the past and how vulnerability to these events might have changed over time.

Society, mobility and bellicosity in northern Mexico

There is a long history of human activity in the northern region of Mexico which corresponds to modern-day Chihuahua, though drought may have influenced the nature of the livelihood strategies developed by people in the region and to some extent constrained the spread of sedentary communities. Archaeological investigations indicate that nomadic hunting and gathering palaeoindian groups may have existed in the region from as early as 14,000 years ago (Lorenzo, 1987; Marquez-Alameda, 1992: 107), and evidence dating from about 10,000 years BP suggests that archaic cultures were still predominantly mobile hunters and gatherers at this stage (Minnis, 1992; Smiley, 1994; Vierra, 1994). By AD 1000, however, those groups that lived along the river valleys, largely as a result of the provision of water for drinking and irrigation, may have begun to develop semi-sedentary livelihoods (Griffen, 1979; Aboites Aguilar, 1994: 15) and by AD 660, the inhabitants of this region had become increasingly dependent upon agricultural production. Indeed, the populations in some parts of northern Mexico at this stage may have been both numerous and culturally quite advanced (Doolittle, 1984), forming what Riley (1976) has termed 'statelets'. Archaeological excavations are suggesting that in some locations, particularly in the north of the region, there may have been some quite substantial sedentary groups reliant on a maize-based subsistence diet, while temporal or rain fed irrigation based semi-sedentary farming communities may have been present in the lower Rio Conchos between AD 1200 and 1400 (Kelley, 1951).

At the time of contact, there were a variety of distinctive cultures in this region. The Chalchihuites settled in a zone to the north of the modern-day state of Zacatecas and around the basins of the Rio Florida, in the southern part of Chihuahua. The Loma San Gabriel Chalchihuites are thought to have occupied this region from the eleventh century. The culture of Loma San Gabriel survived even the collapse of Casas Grandes, the civilization located about 500 km to the north. This was a very advanced and urban civilization, with sophisticated hydraulic technology, irrigated agriculture and water storage. Casas Grandes is thought to have collapsed sometime between 1450 and the early sixteenth century (Cramaussel, 1990a: 84). Whether a period of prolonged drought may have played a role in this remains open to conjecture, though it is known that a severe drought and the dramatic famine known as 1 Rabbit prevailed in central Mexico in the mid-1450s and tree ring evidence points to a period of sustained drought at this time (Hassig, 1981; Therrell et al., 2004).[2]

The Tepehuanes were located in southwest Chihuahua/northern Durango and may have been direct descendants of the Chalchihuites. They practised agricultural production of maize, beans, chile, squash and also some cotton. This group was particularly aggressive, however, dominating other culture groups such as the Acaxees, from whom they exacted tributes of maize and beans, and coming into regular conflict with other groups such as the Tarahumaras. There were three key Tarahumara zones or bands. The first extended from the Fuerte River to the Conchos River, the second from the Conchos to the Papigochic River and the third from there to the less explored territories of the Sierra Madre Occidental. Within these regions, the Tarahumara Indios were very much a single culture but, lacking a central political structure, they lived in dispersed settlements referred to as rancherías (Salmon, 1977).

The majority of the Conchos Basin, from Santa Barbara to the area north of Cuchillo Parado, was inhabited by semi-nomadic groups. The Conchos, or as some of the early Spanish chroniclers referred to them, the Conchería, are thought to have comprised two distinctive linguistic groups: the Sumas-Jumanos of the northern region and the Conchos of the south, occupying the territory of the confluence between the rivers (Conchos and Florido, see figure 2.1). Though little is known about the Conchos ethnographically, they appear to have comprised a large group of people, considering the geographical extent of their territory, and are thought to have lived in small, scattered communities, primarily based on hunting, gathering and fishing, but also practising some agricultural production, cultivating melons, beans, squash and maize (Rodriguez-Chumascado and Espejo expeditions, cited in Hammond and Rey, 1927: 14).

Contact and conquest in the hostile north

Spanish interests in the north of Mexico area initially focused on the potential for mineral wealth. Indeed, only a short time after the conquest of the Aztec capital in central Mexico, rumours began to infiltrate the Spanish conquistadors and first settlers of a legendary region to the north, where there was extraordinary mineral wealth.[3] The Spanish explorers first followed sea-based routes up the Pacific coast in order to explore the area, but the discovery and exploitation of mineral seams in Zacatecas after 1546 provided a base from which land-based explorations of the area further north could proceed (Aboites Aguilar 1994: 13–14). In 1554, Fransisco de Ibarra, who was related to one of the founders of Zacatecas, embarked on a new expedition to discover, and conquer, the mysterious northern kingdom. His expedition paved the way for further Spanish colonization. Durango was founded in 1563 as the capital of the region which became known as Nueva Vizcaya (figure 2.2), incorporating the modern-day states of Sinaloa, Sonora, Durango and Chihuahua, a small portion of the north of Zacatecas and the southwest of Coahuila (Alvarez, 1990: 143; Cramaussel, 1990a, 1990b).

Ibarra undertook additional explorations further north in 1563 and again in 1567 in the hope of further mineral strikes and of discovering the legendary gold rich kingdoms of the north. It was during one of these expeditions that the party encountered, for the first time, what would become the province of Santa Barbara, which for the last thirty years or so of the 1500s would remain a relatively small mining and ranching district. The Rio Conchos provided an important routeway for the Spanish advance into Chihuahua. Towards the close of the sixteenth century, a number of expeditions moved the frontier of Spanish settlement northwards, including those of Chamuscado in 1581, Espejo in 1582–3, Zaldívar in 1588 and Oñate in 1595–8. The Spanish expeditions followed the old routes of the Indios from Durango, passing through Guatimape, Zape and Indé up to Parral and enabled the northward expansion of colonization. A century or so later, the settled area extended up to El Paso, to the northeast including the confluence of the Conchos and Rio Grande Rivers, and to the northwest, towards Casas Grandes (Griffen, 1979).

Early Spanish accounts provide some idea of the size of the native population encountered in the northern part of Mexico at the time of contact. In 1562, for example, Francisco de Ibarra wrote of the 'very populous' Nazas River area of the Tepehuanes, while the Jesuit's Annual letter of 1593 indicates an estimated 100,000 souls occupied the mountains and river regions of the Tepehuan area at that time. Gerhard estimates that the indigenous population of Santa Barbara in 1521 was in the region of 45,000, with a density of around 20 inhabitants per square kilometre. There were thought

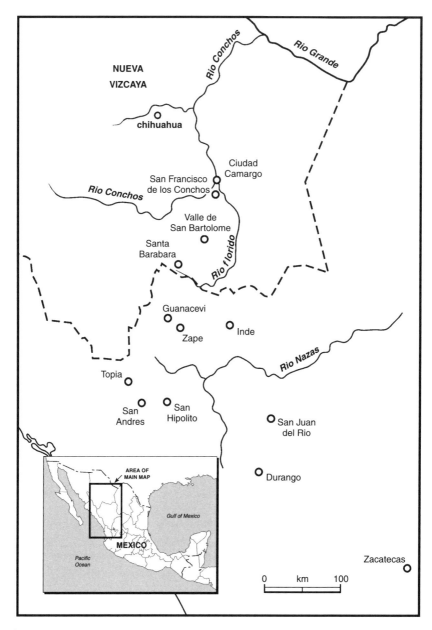

Figure 2.2   Nueva Vizcaya and locations referred to in the text

to be around 1,000 individuals in the Parral region, 3,000 around the Conchos River Basin and 7,000 to the west of the Rio Florido. There were an estimated 10,000 in the vicinity of Inde and 24,000 to the west of the province of Santa Barbara (Cramaussel, 1990a: 86). In total, Deeds (1989) suggests, the Spanish might have encountered a population of around 300,000 semi-sedentary agricultural indigenous peoples in the region.

Spanish colonization proceeded relatively slowly. A rich seam of silver, however, was discovered in Parral in 1631 and further mineral strikes were shortly after made a few miles west in the area that became known as San Diego de Minas Nuevas (Griffen, 1979). This consolidated Spanish interest in the region and stimulated further settlement in Nueva Vizcaya. Parral began to prosper as prospectors descended on the area from as far afield as Mexico City to lay claim to mineral wealth. Indeed, there was, in Griffen's terms, 'a veritable silver rush' to the area. Its population expanded to 1,000 Spanish and 4,000 Indians and Afro-Mexican slaves, the latter brought in as labour for the mines, by 1635. The capital of Nueva Vizcaya was transferred from Durango to Parral, and later San Felipe el Real de Chihuahua (Chihuahua City), following these rich silver strikes, and settlement of the northern area as a whole would be boosted when discoveries of gold and silver were later made near Santa Barbara at San Fransisco del Oro.

Spanish existence in the province, however, was beset by difficulties. This northern frontier of Mexico was bellicose. For the most part, the different indigenous groups of the region lacked a central political structure and were frequently involved in inter-tribal conflict. Nomadic and semi-nomadic bands from the present-day southwest USA, sometimes referred to as Apache (meaning 'enemy') groups frequently raided peoples further south. Various theories have been forwarded to explain these southward migrations, including problems of access to food and water during periods of drought (Tweedie, 1968: 1134). But although conflicts and infractions between different cultural groups were common prior to Spanish contact, there was solidarity in opposition to Spanish colonization of the region. While the war of Mixtón in 1540 was the last great battle between the Spanish and the indigenous groups of the area (Simpson, 1966; Naylor and Polzer, 1986), indigenous rebellions, attacks and thefts by the nomadic hunting and gathering groups remained commonplace throughout the colonial period and into the early years of independence (Aboites Aguilar, 1994: 11).

Perhaps the most significant problem affecting society in the region, however, was and is still drought, and specifically prolonged phases of drought. Tree ring reconstruction of climate changes in Chihuahua (Dias et al., 2002) and northern Mexico in general (Cleaveland et al., 2003) indicates that prolonged periods of consecutive droughts affected the region in the past and contributed to harvest failure, food scarcity and in some cases famine.

Drought also indirectly contributed to economic downturns in the mining industry, affected the health of both the human and animal populations and exacerbated problems of water sharing and distribution between competing users in the agricultural, ranching, mining and domestic arenas. As shall be illustrated in chapter 5, drought may have played a significant role in triggering or at least contributing to some of the reported instances of indigenous unrest and revolt (Florescano et al., 1995; Cleaveland et al., 2003).

Guanajuato and the Chichimec Territory

Location and climatic characteristics of Guanajuato and the Bajío

Guanajuato lies in the Bajío region in the north of the central Mexican highlands (figure 2.3). The Bajío is not well defined physically, but lies between Querétaro, Salvatierra and Leon (Murphy, 1986) and was originally formed during the Tertiary as a vast lake lined by volcanoes. The water drained from the lake to leave a series of interconnected basins lined with lacustrine mud and volcanic ash (Brading, 1978: 13), presenting a potentially very fertile region. The region is bounded by the Sierra de Guanajuato to the north, the Sierra Gorda and the Sierra de Agustinos to the east and the neovolcanic axis, which crosses the country around 19° north, to the south. Los Altos and Jalisco lie to the west.

The area lies on a climatic gradient which varies from semi-arid in the north, to sub-humid in the south, and has a moderately arid climate with a marked pattern of seasonal rainfall. Temperatures vary little between 14° C in January to 22° C in May and the annual precipitation is about 650 mm per annum, with about 80 per cent of the precipitation falling in the summer months between May and October and mostly in the form of afternoon thunderstorms. There are spatial variations in seasonal rainfall, however, with the more southern areas receiving slightly more rainfall than central locations (Brading, 1978: 14). There are very high evapo-transpiration rates, caused by a high humidity, the moderate to high temperatures and high levels of insolation. The region regularly suffers a period of water stress in the hot, dry spring and there are considerable variations both in the intensity and incidence of the summer rains, occasionally resulting in severe droughts.

Much of the region has a very high water table and there are numerous springs and swamps. There are also a number of important rivers in the area, including the Lerma which rises in the Valley of Toluca and enters the Bajío region at Acambaro before flowing onto Salvatierra and Salamanca. The Rio Laja originates in a basin north of San Miguel Allende. The volume of flow in the Laja is a direct function of the amount of rainfall and can thus

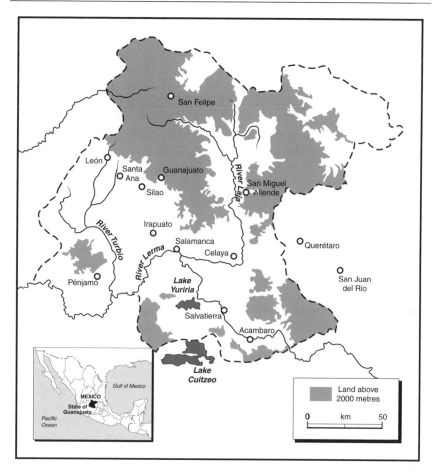

Figure 2.3   Guanajuato and other locations in the Bajío referred to in the text

be prone to seasonal variations in water levels, but the river is also fed by a few modest springs and so has a perennial flow. In addition, there are many ephemeral rivers that form in the hills surrounding the plains of the Bajío in the summer rainy season when water runs off the hills into a series of small streams and arroyos.

Pre-Hispanic settlement and the Chichimec territory

Before the Conquest, the Bajío region represented something of a frontier with only limited settlement. However, there is some evidence of settled occupation in the area prior to the arrival of the Spanish with the first agricultural settlements possibly dating back to between the fifth and first centuries BC. The Chupícuaro culture is thought to have been located in the north of Michoacán and the southern part of Guanajuato and to have flourished between 500 BC and AD 200. This is thought to have been an agricultural society whose communities capitalized on a whole range of lacustrine resources (Wright Carr, 1998: 12–13). Also, during the early centuries of the first millennium, small agricultural communities, associated with the Chalchihuites culture, mentioned earlier, were located along rivers and in defensive locations to the north of the Bajío, into Zactecas and Durango. Around this time in the altiplano regions, however, in the vicinity of present-day San Luis Potosi, there were numerous settlements associated with other communities who practised a mixture of hunting and gathering as well as some agricultural production.

There was a distinctive settlement hierarchy in the valleys of Guanajuato and around modern-day Querétaro. Some large sites are thought to have been regional centres of political control, which were supported by second-ary centres, and had direct control of still smaller settlements responsible for agricultural production and the manufacture of lithic and ceramic objects and tools. The eastern Bajío, including the zones of Dolores Hidalgo, San Miguel de Allende, Celaya, Apaseo, Querétaro and San Juan del Rio, was the seat of power for an important culture, contemporaneous with the classic period in the Valley of Mexico. In fact, this tradition was very differ-ent to that found in the western Bajío, and had more in common with those of the Valley of Mexico, while in the west, culture groups shared similarities with those in the west of Mesoamerica (Wright Carr, 1998: 18).

Between the tenth and the twelfth centuries AD there was a progressive abandonment of many of these agricultural settlements. Some sites may have been subject to a reoccupation, possibly by cultures bearing stylistic similarities to the Toltec, whose primary site at Tula had been abandoned by the twelfth century.[4] Generally, however, resettlement does not appear to have been that common across the Bajío. Indeed, permanent indigenous settlement and land use seems to have been restricted to the southern periphery of the Bajío around Penjamo and Acambaro. Much of this region, however, fell into the hands of diverse, dispersed semi-nomadic tribes, col-lectively referred to as Chichimecas.[5] It was these hostile groups that the Spanish encountered in the Bajío at the time of contact.

European contact and warfare in the Chichimec territory

The greatest Chichimeca 'nation' was that of the Huachichiles, who were one of the largest divisions of the Chichimecas and occupied the region now included in the states of San Luis Potosi, southeastern Coahuila, eastern Zacatecas, northern Guanajuato and eastern Aguascalientes and the southern portions of Nueva Leon (Powell, 1945). Some Huachichile tribes extended further south into the Rio Grande de Santiago. The Guamar, who Las Casas regarded as 'the bravest, most warlike, traitorous and destructive of all the Chichimeca nations and the most astute',[6] occupied the territory around present-day San Miguel, much of Guanajuato and San Felipe, and extended also into Michoacán in west central Mexico, and to the Rio Lerma. Other groups included the Zacatecas Chichimecas further north, the Pamé, whose territory extended as far south as Acambaro in Michoacán, and the Chichimeca Blancos who lived around Aguascalientes, Guanajuato and Querétaro and as far west as the lake district in Jalisco. The Guaxabanas were a group of importance in the sierras of Guanajuato, while the Jonases occupied some areas of Guanajuato and Querétaro. All of these groups, possibly the only exception being the Caxcanes, who established more permanent settlements directly south of modern-day Zacatecas, were seminomadic hunters and gatherers. They shared many customs and subsistence practices, including hunting techniques, though they lacked any central political leadership or legal code and inter-tribal conflict was thus common.

As was the case in the north of the country, the lack of any political unity among indigenous culture groups posed one of the greatest handicaps for the Spanish in this area. The nomadic nature of the enemies rendered it difficult for the Spanish to broach any peace treaties with large segments of the population and thus made these people much more difficult to conquer, relative to the sedentary groups of the Valley of Mexico and those further south (Powell, 1945). In fact, the Chichimecas would be fierce and elusive enemies for four decades after the first Spanish contact in this region.

It should be noted that the Spanish also encountered a sizeable Otomi population in the Bajío region, particularly in the east and north of the Purépecha or Tarascan territory (in Michoacán). These outposts helped to reinforce Spanish defence against the Chichimeca attacks. The Tarascans also employed allied Chichimecas to the same end. The movement of the Otomis into the Bajío proper, however, began shortly after the Spanish conquest of the Aztec capital of Tenochtitlán in order to avoid subjugation to Spanish colonial rule, though ultimately they too would become integrated into the political and economic system of New Spain (Wright Carr, 1998).

The main stimulus for the Spanish to colonize the area came with the discovery of mineral (silver) wealth, firstly, as already noted, north of the region in Zacatecas, and then in the mountains of Guanajuato. These later discoveries stimulated not only the establishment of defensive settlements to protect mineral cargoes against Chichimec raids, but also agricultural investment in the area to support the growing mining communities. As shall be demonstrated in the following chapter, the coupled investments in mining and agriculture helped this area to become not only an important silver mining district, but effectively the agrarian heartland or 'breadbasket' of New Spain's colonial political economy (Martinez de la Rosa, 1965: 41–42).

Power and Political Growth in the Central Valley of Oaxaca

Location and climatic characteristics of the central Valley

Oaxaca lies in the southern highlands of Mexico. Located in the centre of the modern-day state, the Valley of Oaxaca is a drainage basin of 2,500 sq km and represents the largest expanse of flat land in the highlands of southern Mexico (figure 2.4). The elevation of the valley floor varies from 1,420 to 1,740 metres above sea level, with surrounding ridges reaching a maximum height of 3,000 metres (Kirkby, 1973: 7–9). The valley itself is divided by mountain ranges into three distinct regions, with the city of Oaxaca at the centre representing both the administrative and physiographic heartland of the region (Dilley, 1997: 1551). From this centre, the Valley of Etla extends 20 km to the northwest, the Valley of Tlacolula 29 km to the southeast and the Valley of Zimatlán 42 km to the south (Taylor, 1972).

The valley can be demarcated by the Atoyac River Basin. The river itself drains the Etla and Zimatlán valleys, while the Tlacolula arm is fed by one the Atoyac's ephemeral tributaries, the Salado River. The rivers of the region are all characterized by irregular flow, posing a limiting factor for agricultural production (Feinman et al., 1985). Temperatures in the central valleys vary from 11° C in November and December, to 34° C around April and May. Rainfall averages 1,500 mm per annum and, as with all of Mesoamerica, is sharply seasonal, falling in the summer months, between April and October, but the area experiences average seasonal rainfall maxima in June and September. Precipitation in the early part of the growing season (May-June) results from the movement northwards of the ITCZ while easterly winds bring moisture from the Gulf of Mexico. The latter part of the wet season (July-September) is governed by the Mexican monsoon, but there is a relatively drier period between July and August, known as the sequia intraestival or mid-season drought, which varies in intensity and exact timing from year to year. Each arm of the valley, however, is characterized by a distinctive geography with different topographic and hydrographic conditions and there is considerable spatial variability in climatic characteristics across the three areas (Dilley, 1997).

There are three primary topographic zones in the valley area: an alluvial plain, a gently sloping piedmont and a more steeply sloping mountain area.

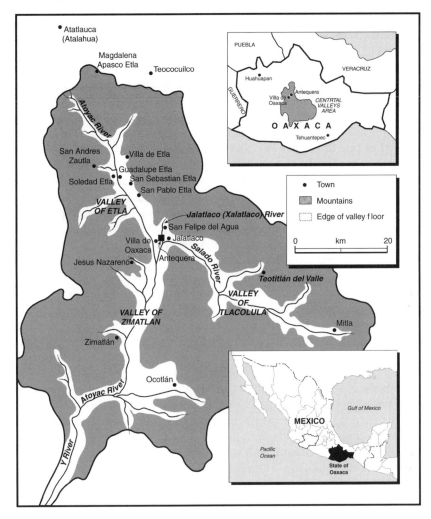

Figure 2.4   The Valley of Oaxaca (incorporating the Etla, Tlacolula and Zimatlán valleys) and locations referred to in the text

Most agricultural activity takes place in the first two of these zones (Sanders and Nichols, 1988). As Kirkby (1973) has shown, the annual precipitation within the zone of cultivation is moderate. Adequate soil moisture constitutes a critical problem as evaporation rates can exceed precipitation in some parts of the valley. Indeed, crop losses due to a lack of adequate moisture are common.

Political growth and decentralization in pre-Columbian Oaxaca

The flat fertile alluvial soils of the central valleys, and the availability of both permanent and temporary water sources, provided an attractive base for both pre- and post-Hispanic society. The Valley of Oaxaca has a long history as a major power centre in the southern highlands of Mexico, and has been included as one of the five key areas of Mesoamerica which are thought to have been instrumental in cultural evolution over wide geographical regions (Palerm and Wolf, 1957). The valleys have been continuously inhabited from c. 8000 years BC. Settlement is first documented during the Archaic period (8000–1800 BC), though population densities during this time are thought to have been relatively low (Flannery and Spores, 1983), a reflection perhaps of the apparently dry climatic conditions which may have prevailed during this time. Certainly this would have limited the opportunities for agricultural production (Joyce and Mueller, 1997). Between 1800 and 900 BC there are thought to have been significant changes in land use, with the development of the first sedentary villages in the valleys (Flannery, 1969). To some extent, these developments may have been a function of a shift towards wetter conditions, an increased growing season and greater crop productivity, allowing an increase in population. By 1000 BC, separate polities had developed in each of the three arms of the Valley of Oaxaca (Flannery et al., 1981; Kowalewski et al., 1989; Ballansky, 1998), though at this stage the majority of the population (estimated to be 52 per cent) was based in the Etla arm (Feinman et al., 1985). The southern arm of the valley was relatively sparsely populated, despite its apparently fertile soil, suggesting perhaps that the availability of good farming land may have been only one consideration in the location of pre-Hispanic Oaxacan villages (Feinman et al., 1985).

There was further demographic and political growth in the region between 700–300 BC, though the expansion in population and the greater investment in agriculture is also reflected in a possible increase in anthropogenic soil erosion during this period (Joyce and Mueller, 1997). It is interesting to note, however, that most of the population continued to live in small settlements (no more than 3 ha in size) and much potentially productive land remained unoccupied (Feinman et al., 1985). Nonetheless, between about 500 BC and AD 1 the population of the valley area as a whole may have increased 28 fold (Nicholas et al., 1986). Consequently, there was a dramatic increase in settlement and one of Mesoamerica's earliest cities, Monte Alban, which covered 442 ha by this stage and housed between 10,200 and 20,400 people, flourished (Joyce and Mueller, 1997).

Developments in agrarian practice might have helped stimulate the development of Monte Alban as a political and administrative centre.

The city's growth, for example, has been attributed to the capacity for agricultural development in the area and specifically the ability to trap floodwaters for irrigation purposes which could support intensive agriculture. There was also an extensive civic water drainage system in the city (O'Brien et al., 1980). In fact, small-scale irrigation was probably used throughout the valley area to produce two annual harvests of maize (Kowalewski et al., 1989). Archaeological surveys near the village of Xoxocotlán on the southern piedmont slopes of Monte Alban, for example, have revealed evidence of an irrigation system, a dam and a 2 km canal used for this purpose (Neely, 1972; Neely and O'Brien, 1973).

The rise of Monte Alban coincided with a growth in public and ceremonial buildings at many other sites and there were numerous secondary centres of the city in each arm of the Valley of Oaxaca (Marcus and Flannery, 1996; Kowalewski et al., 1989). There was a marked increase in political and religious activities across the region and population growth is thought to have been in the region of 1.4 per cent a year. This growth may have reflected some degree of in-migration, though rapid population increases may have also have been a function of changes in household strategies and fertility. Rising tribute demands, for example, might have encouraged more children in each family to facilitate more intensive production of tribute (taxable) products (Feinman et al., 1985).

While Monte Alban retained is unparalleled political and administrative importance in the region, by 200 BC there had been some decentralization. New settlements began to emerge in defendable locations, possibly suggesting some level of tension between local districts. By the Early Classic period (AD 200–450), there is evidence to suggest there were other important population centres in the valley which became much better integrated both politically and economically (Feinman et al., 1985). Although valley-wide political integration had long ensured that food deficits could be met by transporting goods produced elsewhere in the region (Sanders and Nichols, 1988), exchange networks were much better developed by this stage. Periodic surpluses might have been used to support drought afflicted areas, as well as feeding the urban population of Monte Alban (Feinman et al., 1985).[7] As shall be illustrated in chapter 4, such networks and exchanges continued to operate after the Conquest and were pivotal coping mechanisms in Oaxaca and elsewhere during periods of harvest crisis.

Settlements on defendable hilltop locations had begun to emerge in the southern end of the valley by AD 450, though by this stage the population of the Etla valley had begun to decline. However, some time around AD 600–700 there was an increase in the amount of land that was exploited, including areas classed as agriculturally marginal or high risk (Sanders and Nichols, 1988). The population of the valley increased almost threefold during Monte Alban's period of supremacy (Feinman et al., 1985). When the city reached its maximum population of 15,000–30,000 sometime after about AD 500, it is thought that population pressure may have begun to exceed the supporting capacity of its surrounding area. Greater exploitation of less favourable areas of land accelerated soil erosion and this in turn may have contributed to a period of cultural collapse known as the Epiclassic demographic decline in the Valley of Oaxaca about AD 750 (Balkansky et al., 2000: 368). Population began to decline and settlements were dramatically depopulated, including those at distances of up to 20 kilometres from Monte Alban (Feinman et al., 1985). There is considerable debate over the nature of the political developments immediately following this period of collapse (Marcus and Flannery, 1996; Winter, 1989), though it is thought that many of the urban centres faced decline and that political decentralization and continued population loss followed (Spores, 1972; Kowalewski et al., 1989). Administration in the valley is thought to have disintegrated into many different political fragments, governed by indigenous elites (Balkansky et al., 2000).

As a consequence of the devolution and decentralization resulting from the dissolving Monte Alban state, the first Spaniards in the area encountered multiple polities. The Valley of Oaxaca was politically organized into a number of nucleated town states or communities. Each community was governed by a local dynasty or hereditary aristocracy headed by a cacique. Between 1000 and 1500, there was an increase in cacicazgos (the inheritance and lands of the caciques). Each cacique had virtually absolute power within their community, but political organization above the community level was relatively weak in the valley area. There were some loose federations, dominated by a lord (principal), of whom the cacique was supreme lord. Tribute was exacted from subject populations by the supreme lord. Succession to this office was hereditary, a pattern which is thought to have prevailed in the Mixteca Alta (Spores, 1967, 1972). At the other end of the social spectrum were the macehuales, or free commoners, the mayeques, or serfs, and the slaves. Macehuales formed the majority of the population and consisted mainly of farmers, but also artisans, shamans and some merchants (Chance, 1976).

Ethnic diversity and organization at the time of contact

Arriving in Oaxaca from Veracruz in 1519, the Spanish conquistador Hernando Cortes and his troops encountered one of the most ethnically and linguistically diverse regions of Mesoamerica (Chance, 1976). There are thought to have been a minimum of 16 different languages spoken in the region (Christensen, 1998), though the development of this sociocultural complexity has been a source of academic contention (Zeitlin, 1990).

By the time Cortes's Leiutenant, Pedro de Alvarado, first entered the southern Isthmus in 1522, he encountered a complex cultural landscape that reflected the turbulent political history of the region. Recent investigations using primordial titles[8] have demonstrated competing interpretations of the Spanish conquest of Oaxaca from both Mixtec and Aztec perspectives (Sousa and Terraciano, 2003), but it is clear that at the time of European contact, the area was densely populated by three culture groups, the Mixtecs, Zapotecs and Aztecs.[9] Zapotecs occupied the central valleys, though they had also secured an important base from which to expand on the coastal plain at the site of Tehuantepec (Zeitlin, 1990: 32), while the more rugged, mountainous land in northern and western Oaxaca was home to the Mixtecs. By 1350, however, the Mixtecs had begun to infiltrate the Valley of Oaxaca. Although the exact date of their arrival is open to debate, by the 1480s, the Aztecs has also begun to settle in the valleys, with the goal of subjecting the existing Mixtec and Zapotec populations to tribute. They established a garrison at Huaxyacac, which would later be the location of the capital city of the region, Antequera (now Oaxaca City).[10]

Like other regions in Mesoamerica, maize, beans and chile represented the staple crops grown in the area, though there were in fact many different cultivars (Taylor, 1978: 74). The caciques and principales held a monopoly of the most fertile and best-watered territory and tribute-paying macehuales cultivated both communal lands to produce tribute commodities and small plots for subsistence purposes.[11] The mayeques (referred to as terrasguerros after the conquest) (Taylor, 1972) were bound to the cacique lands (and possibly also those of the principales) and belonged to the caciques' inheritance. The slaves were at the bottom of the pre-Hispanic social hierarchy. They were usually obtained through warfare and were sold at markets, such as that at Miahuatlán.

Estimates suggest an indigenous population of 350,000 in the valley on the eve of conquest (Cook and Borah, 1960, 1968; Taylor, 1972), of whom 78 per cent were Zapotec, 10–20 per cent were Mixtec and 1–2 per cent were Nahua or Aztec (Chance, 1976). The valley's Zapotec and Mixtec populations submitted relatively peacefully to the Spanish conquistadores in 1521. As a reward, caciques from both groups received special grants from the Spanish and in return for their support were offered some degree of protection (Taylor, 1972). Indeed, as shall be discussed, because of this protection, the cacicazgos would remain in control of their land and property until well into the colonial period. This indigenous land retention would have implications not only for the maintenance of a strong indigenous identity in this region, but also for the development of a distinctive colonial political economy. Both factors very much influenced the vulnerability of society in this region to climate variability.

Chapter Three

Exploring the Anatomy of Vulnerability in Colonial Mexico

Introduction

The environmental, geographical and cultural differences between and also within the three areas of concern at the time of contact fostered the evolution of discrete regional socioeconomic systems and colonial political economies. The specific social, economic, demographic and environmental contexts were pivotal to shaping the level of vulnerability to climate variability and weather events in each case study area, and may have also influenced the nature of the resultant impacts and the relative success or failure of responsive and adaptive strategies. For this reason, it is important to consider the general mechanisms and instruments that facilitated Spanish colonization of the country as a whole, and also the regionally distinctive developments that took place in each case study area.

The Tools of Conquest and Colonization

Land and labour in early colonial Mexico

A set of policies concerning colonization and land acquisition, use and tenure operative in Spain on the eve of Conquest, or devised immediately afterwards, was transferred to this part of the New World. The Crown divided society in Spanish America into Republicas de Españoles and Republicas de Indios. Each Indio republic was presided over by an Indio governor or mayor, usually a member of the indigenous nobility. In return for juridical protection and agricultural benefits, the Crown collected an annual tithe or tribute from indigenous subjects. Normally, this obligation fell to the male head of family – a system of classification and tribute exaction that would remain unchanged until the middle of the eighteenth century.[1]

Royal decrees, compiled and issued between 1530 and 1550, established a system of hierarchically ordered settlements across New Spain (Licate, 1981). The territory was first divided into major provinces (provincias mayor), within which existed minor provinces (provincias menor or partidos), also known as the Alcadía Mayor and Corregimiento, which in turn encompassed a series of new or existing Indio towns (pueblos). Scattered indigenous communities were brought together and relocated (reducción) in a series of new towns, or congregaciones, with the dual purpose of imposing control and converting the Indio populations to Christianity (Licate, 1981), and all towns were organized in a system of cabeceras or head towns, and sujetos – their dependencies. The cabecera was the means through which the Spanish organized tribute and labour for their own use, while drawing on the subject villages in order to meet the tribute and labour quotas. Spanish officials entered the region as Corregidores (officials), Regidores (councillors) lieutenants and constables, and as Alcades Ordinarios (magistrates).

A series of municipios were also established representing small jurisdictional units delimited by the districts of a town or a city and the lowest level of the Spanish American administration (Bacigalupo, 1981). These units were governed by an ayuntamiento, or a council (cabildo), composed of six or twelve Regidores, according to the importance of the locality, and two Alcades. It was such people who held the vara de justicia (wand of justice) for particular districts, handling complaints, including pleitos (disputes) over land, water and property, and implementing Crown legislation and policies.

Newly conquered lands, and the environmental resources within those lands, immediately became the property of the Crown. The law of the first conquistadores, for example, stated that 'the water belonged to the king, just as did the lands, the fields, the forests and the pastures' (cited in Solórzano y Pereyra, 1776), although legislation instituted in the early colonial period decreed that 'the Indios should be left their lands, their inheritance and fields and pastures ... and they should not have a want of that which is necessary, and they have all the help and support possible for the sustenance of their houses and families' (cited in Florescano, 1976). In theory, the Crown provided each Indio republic with institutional arrangements 'most suited to its needs' (Cope, 1994: 3). Some of this vast and unexplored 'new' territory, however, was also partitioned among the conquistadores as rewards for their role in the Conquest. Each conquistador was entitled to two caballerías[2] of agricultural land to cultivate (Florescano, 1976), while more substantial royal 'gifts' were presented to certain members of the conquering forces. The encomienda system represented one such recompense, being at first imposed on the indigenous population as a form of repartimiento[3] grant to conquistadores, but also in an effort to reconcile the conflicting religious and capitalist purposes underlying Spain's ideals for this part of their New World. In return for protecting the Indios, if only theoretically, and instructing them in the Catholic faith, the encomendero or landlord of the encomienda could exact tribute from them in the form of goods, money and, in the case of the first encomienda grants, labour (Butzer, 1991).

Overall control of the lands and peoples of the new colony, however, was the charge of the first Audiencia of Mexico. Established in 1528, the Audiencia represented the main administrative council for New Spain next to the Spanish Crown. Under the auspices of this first corrupt administration, however, several unofficial encomienda grants were awarded. It also became the norm for encomenderos to abuse the tribute system, exacting excessive demands from their subjects (Paredes, 1979). Overexploitation of this kind was to become particularly severe with the expansion of mining activities in the colony.[4] In 1532, however, a new type of reformed encomienda (Ugarte, 1992: 209), based on an idea initially forwarded by Bishop Juan Ramirez de Fuenleal, president of the second Audiencia, came into operation.[5] Encomienda privileges were curtailed and more restrictive controls on the use of labour were introduced in the 1540s (Prem, 1992). Tribute taken from Indios was fixed and regulated, while Indio enslavement, even as punishment, became prohibited. In 1629, further legislation was issued to finally render encomienda grants inoperative after their so-called 'fifth life', that is to say after five generations of existence. Finally, in 1718, most encomiendas in the Spanish colonial empire had been abolished, bar some in parts of the Yucatan, Chile and in Paraguay (Ugarte, 1992) and as shall be illustrated, also in the north of Mexico (Deeds, 1989).

The establishment of a more formal legislative system under the second Audiencia led to a review of the land granting process, and a series of new regulations was introduced.[6] From 1536, land could only pass into private possession by means of a royal award or grant (merced real) issued and confirmed by the king (Florescano, 1976). Mercedes reales began to be granted officially by the Viceroy of New Spain in 1542. The measures for each grant depended upon whether the land was to be used for livestock grazing (estancias) or agriculture (caballerías de tierra). Each caballería is thought to have measured about 42.8 hectares (or 106 acres), although there is some contention over the exact figure (Butzer and Butzer, 1995). The area of land granted to keep livestock depended on the nature of livestock to be kept. Sitios (later referred to as estancias) for pigs, sheep or goats (ganado menor) measured two thirds of a legua (league) square (2,793 by 2,793 metres, 780 hectares or 1,927 acres), while cattle (or horses and mares) were kept on plots (estancias de ganado mayor, sometimes written 'maior') measuring a legua square (4,190 by 4,190 metres, 1,756 hectares or 4,336 acres).[7] These measures were not, however, formally specified and ratified until 1560 (Butzer and Butzer, 1995).

Strict regulations accompanied the granting of lands, designed to mollify the dual problems of overgrazing and crop depredation, and reflective of a long history of both in Spain. It was deemed that no livestock other than draft animals could be kept on agricultural lands, though the legislation enforcing this was not officially passed until 1581.[8] There were also fixed stocking rates on the estancias,[9] with limits imposed on the amount of livestock held within a delimited area being issued in 1564. With the award of a grant also came an agreement to make certain improvements to the lands in question. Stipulations for estancia awards, for example, included the construction of barns and the building of livestock corrals to prevent depredation (Butzer and Butzer, 1993). In addition, within one year of the grant being made, lands had to be ploughed and sown if they had been granted for agricultural purposes, and 'populated' with cattle/horses, sheep or goats if the land was designated for livestock.[10] If the grant had been awarded for agricultural purposes, harvested fields had to be made available for pasturage. A grantee could not sell-on the merced within four years of the award being made and it was forbidden to pass or sell land into the hands of 'the church, a monastery, a hospital or anyone ecclesiastical' (Florescano, 1976), each award enforcing this regulation in writing. The latter condition was added no doubt because the Church had been considered an institution of privilege and corruption in fifteenth-century Spain, and represented a powerful and influential threat to the authority of the Spanish Crown (Gibson, 1964).

There was a certain amount of protection afforded Indio communities and their territories, at least in the early post-Conquest period. Mercedes awards stipulated that the land use for which the grant was being awarded would proceed 'sin perjuicio de tercero', that is to say, without any 'injury' or harm being inflicted on any third party.[11] Indeed, land grants could be declared void if negligence could be proven or if any environmental deterioration was seen to have resulted after the award was made (Florescano, 1976). Special legal requirements also ensured, at least theoretically, that land grant awards would not impinge on any neighbouring indigenous property. Indio lands per se were effectively sacrosanct as long as they remained cultivated, and were declared inalienable except under special circumstances, with the approval of the Viceroy in office at the time, while 'buffer zones' were created between Indio and Spanish lands in order to protect the Indio possessions from Spanish usurpation. During the administration of the second Viceroy, Don Luis de Velasco (I) (1550–1564), for example, estancias awarded to Spaniards had to be at least a league (4,190 m) away from the nearest Indio lands and at least half a league away from the nearest cultivated territory. In the regulations of 1567, however, these measures were reduced to 1,000 varas (837 m) and 500 varas (418.5 m), respectively.

In addition to these protective measures, provision was made to ensure indigenous communities had sufficient lands for the production of subsistence and tribute goods. Each indigenous settlement had sole use of the lands up to a distance of 500 varas in each cardinal direction from the centre of town, the measures increasing to 600 varas/502 m in 1687.[12] Indigenous communities were also entitled to an ejido (area of common land) measuring one square league (1,750 hectares) according to a Real cédula of 1573 (reaffirmed in 1713) in which there was designated communal access for timber cutting, grazing livestock, hunting animals and gathering various products of the forest. The 'legal fund', as the 600 varas distance came to be known (Barrett, 1973), and the ejido were intended to be the minimum lands to which the communities held rights, and as the decree made plain, they were to have whatever additional lands they required.[13]

Indio communities generally also enjoyed favourable legal status with respect to water rights. A royal decree issued in November 1536 illustrates the Crown's commitment to maintaining the traditional indigenous systems of resource exploitation, suggesting that 'the distribution of waters should be according to Indio custom', while indigenous water rights were generally defined by the needs of the community. As shall be considered in chapter 5, this concept could be used to limit as well as legitimate water rights and on a number of occasions led to conflicts over water rights. Water management and access also featured among the 54 articles of the New Laws of the Indies issued in 1542 (Simpson, 1966). 'The use of water was to be communal … we order that the use of all the pastures, woodlands, waters of the provinces of the Indies are to be common to all the present residents, and to those of the future so that they can enjoy them freely' (cited in Musset, 1992).[14] The right of common use of water, however, was understood to apply only to the use of streams and rivers for drinking, fishing and domestic purposes. It did not permit the free diversion of water for irrigation and forbade unauthorized private appropriations (Murphy, 1986). Communities, institutions or individual landowners could, however, petition for a merced of water. Early grants for water were, for the most part, awarded relatively straightforwardly in return for cash payments of variable amounts (Lipsett-Rivera, 1999b: 27), though generally, mercedes for water rights per se were rare (Prem, 1974; Murphy, 1986). Land grants could be awarded, however, with water rights included. Murphy (1986) highlights three different forms of water concession. These included awards made of caballerías de riego, or irrigable land, which in effect were construed to confer water rights. Land might also be awarded 'con el agua necessario', that is to say, with the necessary water needed for a specified use. As long as the conditions that accompanied a merced grant had been met, such a grant also provided a right to any water within the land. Water that was not subject to any form of merced remained the property of the Crown according to royal decrees. There was, therefore, a suite of legislative procedures and environmental management systems in place to ensure at least theoretical protection of indigenous land and water rights. In reality, however, the territorial and water related privileges afforded the indigenous communities would be ignored and their livelihoods thereby threatened.

The destruction of 'Utopia': indigenous depopulation and the emergence of monopolies

It has generally been accepted that 'virgin soil epidemics' – the introduction of Old World diseases to a people and land with little or no resistance (Crosby, 1979, 1986) – triggered massive depopulation across the country throughout the sixteenth and early seventeenth centuries (Roberts, 1989). Waves of epidemic disease swept across Mexico causing unprecedented loss of life among indigenous populations (Prem, 1992). Depopulation was especially severe in the lowlands and coastal areas where high temperatures and humid conditions favoured the transmission of pathogens, trends which have been recognized elsewhere in the Americas subsequent to European contact (e.g. Newson, 1985; Villamarín and Villamarín, 1992). There is still debate as to the actual size of the pre-Columbian populations and the magnitude and scale of their decline (Roberts, 1989),[15] and medical historians have long speculated on the actual nature of the disease(s) involved in individual epidemics (Carter, 1931; Nicolle, 1933; Zinsser, 1935), suggesting that smallpox, yellow fever, measles, plague, chicken pox and typhus may have all been significant, while Gibson (1964) notes that there were several introduced diseases that appeared in multiple episodes, or what Borah (1992: 7) has termed 'compound epidemics'.[16]

While there are many localized instances of disease reported in the archives throughout the colonial period, there were several major epidemic periods in 1520, 1531, 1545 and 1575–6 which resulted in a series of abrupt, irregular phases of depopulation (Slicher van Bath, 1978; Whitmore, 1991). These epidemics are thought to have been separated by phases of rapid recovery – a consequence perhaps of the high fertility rates that prevailed among native populations. Before a population could rebound demographically, however, another epidemic usually struck so that the 'die off' became cumulative, eventually leading to demographic collapse (Whitmore, 1991). Ironically, the Spanish colonial settlement policies of reducción and congregación of residual indigenous populations facilitated the spread of the contagion and disease may have diffused along trade routes, with no direct

Spanish contact necessarily being implicated in the transmission process (Gerhard, 1982).

It has recently been suggested that some of the most serious epidemics to strike central Mexico in the sixteenth century may have been caused by the same haemorrhagic fever (Acuña-Soto et al., 2000). The disease was sufficiently different to any earlier or subsequent instances recognized as yellow fever, typhus or smallpox. Perhaps most controversially, but in accordance with the suggestions of one of the earliest medical scholars in this field (Zinsser, 1935), it has been posited that this fever may have had a New World etiological agent (Marr and Kiracofe, 2000). Moreover, the same fever is thought to have reappeared periodically throughout the colonial period. Links have been drawn between extreme climate events, dramatic changes to social structures and agrarian practices such as those associated with conquest, and the recurrence of such disease epidemics (Acuña-Soto et al., 2002). As suggested in chapter 1, extreme weather events may certainly influence the timing and intensity of disease outbreaks (Epstein, 1999, 2001; Patz et al., 2005), while both malnutrition and famine conditions consequent upon weather related harvest losses can exacerbate and facilitate the spread of infectious diseases (Kovats et al., 2003). Moreover, when a major upheaval occurs owing to drought, flooding or perhaps other natural disasters such as earthquakes, populations can be dispersed, mixing with others who have less immunity and a new phase of epidemic disease may emerge with 'horrifying rapidity' (Burroughs, 2001: 13). There is some evidence from medical archives and tree ring records to suggest that drought, or drought punctuated by heavy rainfall and flooding, may have interacted with prevalent ecological and sociological conditions, magnifying the impact of infectious disease in central Mexico (and Mesoamerica more generally) and may go some way to explaining at least some of the periods of epidemic disease (Acuña-Soto et al., 2000, 2002, 2004, 2005). This hypothesis will be considered further in chapters 5 and 6.

Whatever the cause, in many parts of the country, the coupled impacts of depopulation and the land grant policies and tenure systems that came into operation in the first decades of the colonial regime contributed to the progressive alienation of indigenous communities from their lands or realms, their resources and their customary resource management strategies. Land granting in some regions peaked following the most serious epidemics of the sixteenth century as depopulation led to the abandonment and vacation of lands, which then became open to appropriation.

Two policies issued between 1591 and 1616 were to assist in the transfer of territorial possession and land use from Indio to Spanish hands, and contributed to the undermining of indigenous rights to access land and water resources. Mercedes became open to appropriation by auction and effectively began to be sold to the highest bidder.[17] As depopulation rendered residual indigenous communities less self-sufficient and unable to meet tribute demands, it was not unusual for individual Indios to sell their former lands at ridiculously low prices as mercedes to landowners in order to earn some immediate income. In other cases, direct purchases took place so that the Spanish landowners could bypass the often lengthy legislative process that was undertaken to avoid granting land that was already under previous title or claim (Prem, 1992).

An allied land grant policy, referred to as composición, was to assist in these processes of progressive alienation and agglomeration of territory. Gradually replacing the merced, which was phased out after 1618 and suspended more formally as a policy in 1643 (Florescano, 1976), composición was a process whereby title to lands could be confirmed and regularized by a small monetary payment. In this way, the Spanish Crown effectively sanctioned individuals and institutions to appropriate lands that had been designated indigenous territory, or communal pasture and woodland, according to both pre- and early post-Hispanic legislation, and thus legalized usurpation and illegal 'invasions' of former Indio territory. Furthermore, landowners with defective titles to water could also use this policy to legitimize possession of water as well as land, although officially composición could not be used to legitimize any lands and water resources taken unlawfully from Indios.

Particular landed families also often secured their possessions through a feudal device known as mayorazgo or entail. Following the owner's demonstration that an estate was valuable enough to justify entailment and when the required fees had been paid, the Crown approved the formation of such mayorazgos. This institution effectively prevented heirs from selling or dividing the entailed property, and thus it passed intact from generation to generation (Burkholder and Johnson, 1994).

Although merced regulations forbade direct awards to any 'persona ecclesiastica', all forms of pious donation were permitted, allowing the Church to expand the territory under its control. To begin with many members of the first religious orders to arrive in New Spain managed to accrue lands via donation from indigenous communities and from wealthy Spanish landowners. Capellanías (grants, usually of money and more rarely of land, given to individual clerics or groups for religious and rogation ceremonies) were also an important means of acquiring rural land for the more urban convents and monasteries in colonial Mexico (Taylor, 1972: 168; Brading, 1978). Time and again religious orders would also obtain titles to land through individual purchasers acting in proxy for the Church, while 'licencias para sembrar', permits to sow land, could also be granted for livestock oriented estancias, thus converting a sizeable area of grazing land over to cultivation (Prem, 1992). The scale and impact of donations, licenses and purchases were such that the profits of the Church were invested in the construction of innumerable monasteries, churches, chapels, colleges and religious buildings. Various religious institutions also invested in private mills, plantations, estates and livestock estancias. Donation of lands and financial backing allowed the Jesuits in particular to become among the most powerful of the landowners in New Spain. It has been suggested that half of New Spain was under Jesuit control by 1599 (Ugarte, 1992: 231).

By the close of the eighteenth century there were an estimated 10,433 haciendas and ranches in New Spain. Between 1640 and 1700 individual landowners across many parts of the country amassed vast tracts of land via official merced awards, direct purchase and usurpation, while most of the emergent agricultural haciendas, livestock latifundias and the great properties of the Church had been legalized and confirmed by composición. These landholdings incorporated formerly indigenous territory and in many cases this restricted access to the most basic of natural resources for survival: water and land. Indeed, as land became a more sought after commodity, the indigenous populations were to lose whatever privileges they initially held in terms of access to and control of their lands and natural resources. The provision of an ejido and the 1,100 vara buffer zone began to be ignored, and only the 600 vara legal fund was to remain in force in the later eighteenth century.[18] Indio communities and individuals actively attempted to defend the land and water against such usurpation by similarly seeking to obtain rights to land through composición, but they often lacked specific titles and claimed land on the basis of 'possession immemorial'. This latter claim was valid under a concept known as prescripción, but was more easily challenged than written documents of title. In this respect, indigenous land loss was inevitable. As shall be illustrated, however, some indigenous communities and particularly those in Oaxaca were rather more successful in defending their lands, rights and resources.

The Emergence of Regional Colonial Political Economies

Pacifying and settling the northern frontier

Far removed from the controlling influence of the capital and with a numerically weak European population in the early years of the colony, fears of indigenous attacks were ever present in northern New Spain and fuelled an almost constant Spanish military presence (Santiago, 1996). Exploitation of the mines in the region was thus accompanied by administrative problems. The need for protection of the silver frontier forced the Spaniards to develop military methods somewhat different from those which had been employed in the Conquest, while the guerrilla tactics employed by many of the more nomadic indigenous groups required more substantive and strategic Spanish defensive co-ordination. This involved the development of presidios and towns placed at key points along the frontier of northern New Spain. A chain of forts had been established across the northern frontier by the close of the seventeenth century (Griffen, 1979).

Various lists indicate that Conchos and Tobosos Indios were regularly enlisted by the Spanish as auxiliary troops in the defence of the northern frontier against more antagonistic raiding tribes, a practice that was to play an important role in the process of indigenous acculturation. For the most part, however, the task of 'civilizing' the semi-nomadic and hostile groups in the north of Mexico through conversion to Christianity, and hence paving the way for Spanish colonization, fell to the Franciscan and, a little later, Jesuit orders who established missions across the region. The first missionaries ventured into the region in about 1531, encountering both placid as well as violent responses from local indigenous groups (Gradie, 2000: 95). It was not until later in the sixteenth century and during the seventeenth century, however, that missionary activities expanded. It was this mission enterprise that led to the congregación of the different Conchos, Acaxee, Tepehuan and Tarahumara groups (Deeds, 1989). In 1574, the Franciscans received a royal decree which authorized the founding of convents in Durango, San Juan and Santa Barbara,[19] and in 1604, they founded the convent of San Francisco de Conchos, about 70 km to the northeast of Santa Barbara. This was the first mission established specifically to congregate and evangelize the Conchos (Aboites Aguilar, 1995), but following the settlement of Parral, missionary activity increased across Nueva Vizcaya. As well as providing a means of pacifying the hostilities, these congregaciones paved the way for a ready supply of indigenous labour for the expanding Spanish mining and agricultural activities in the region (Deeds, 1989).

The mission came to represent one of the principal contact institutions on the northern frontier of New Spain and missionaries were in part responsible for introducing agricultural practices, including crop rotation and irrigation, to the more semi-nomadic Indios. The Franciscans were concerned with the congregaciones of the Tepehuanes and Conchos in the Santa Barbara and San Bartolomé area, while the Jesuits tended to focus their activities on the western edge of the region (Deeds, 1989). Both missionary enterprises, however, were retarded by the attacks by the nomadic indigenous populations across the region that they were attempting to subdue, many of which, as shall be considered in chapter 5, coincided with periods of dearth and disease across the north of the country. Moreover, Griffen (1979: 51) notes that the endeavours of the Franciscan fathers were subject to the 'hindrance and interference from other sectors of Spanish society'. Indios were often induced away to work in the mines or on local haciendas.

There was a great need for labourers for the burgeoning mining economy and the emergent agricultural economy in this region and labour supply certainly posed a constant problem for the Spanish colonists and prospectors in this region (Griffen, 1979; Cramaussel, 1990a: 33).[20] The earliest labour drafts were drawn from Indio rancherias rather than the more sedentary pueblos, and as Deeds (1989: 433) notes, Indio 'middlemen' might have often facilitated recruitment. The first concession referred to as an encomienda was to a vecino (or resident/citizen) of the Santa Barbara province in 1577, when Martin Lopez de Ibarra, Francisco's cousin and a government officer at this time, awarded a concession of Conchos to Pedro Sanchez de Fuensalida (Cramaussel, 1990a: 41).

The Conchos in fact played a key role in the development of the province following conquest and were regarded as a relatively steady supply of labour in Parral, in Santa Barbara and the San Bartolome Valley.[21] Labour was also drafted from Indio groups taken in warfare (Griffen, 1979: 46), [22] while Indios were also being pressed into service in the sierras, where silver strikes helped stimulate the establishment of settlements at Guancevi, Topia, San Andrés and San Hipólito. Other groups were also drafted from even further afield.[23] The distances between Spanish settlements and the locations of encomienda grants could be up to hundreds of miles, leading to permanent (re)settlement of Indios in some situations. Literally hundreds of Indios were brought to the Parral district from the west coast and from New Mexico during the 1600s (West, 1949; Griffen, 1979: 46–48). Despite being officially prohibited in the region by a Real cédula issued by then Governor Francisco de Urdinola in 1609, therefore, and notwithstanding continued requests for the practice of labour drafts to cease, the encomienda and repartimiento systems appear to have prevailed in this part of Mexico at least into the eighteenth century (Bancroft, 1884: 586; Griffen 1979: 4; Deeds, 1989).

Moreover, repartimiento abuses appear to have been a persistent problem, particularly for mission towns. In 1746, when the new mission at Batopilas was established, for example, the Viceroy and Captain Don Juan Fransisco de Guemes y Horcasitas expressed his concern in this respect, suggesting that 'the abuse of the repartimientos of Indios in Nueva Vizcaya ... has left many missions and towns deserted', which in turn had led to a shortage of labour for cultivation such that a number of towns and missions were unable to sustain themselves.[24] The Viceroy introduced strict regulations, declaring that Indio groups drawn from the region were 'not able to be pressed into service and personal work in the first ten years',[25] and that mining impresarios should be prevented from retaining Indios in their service for more than a month. Drafts were not supposed to involve the removal of individuals more than 10 leagues from the town, and finally, the Indios and missionaries were required to provide an annual report on the number of individuals that resided in the mission towns, 'their status and condition'.[26] By 1751, the Corregidor of San Felipe el Real (Chihuahua), Don Antonio Guitierrez de Noriega, had received a letter despatched from the superior government detailing these regulations and instructing him to levy charges and penalties (to the tune of 200 pesos) against anyone contravening them.[27] Abuses continued, however, and people did leave mission towns. In fact, the loss of Indios to repartimiento labour drafts remained problematic throughout the colonial period.

Agrarian development in the hostile north

The burgeoning mining economy in northern New Spain was supported by greater Spanish investment in agriculture. This investment was also a function of the rapid indigenous depopulation across the northern region in the sixteenth and early seventeenth centuries. This has commonly been attributed to the introduction of new pathogens, but was most probably also exacerbated by the harsh working conditions, particularly in the mines,[28] a situation which in turn is thought to have contributed to some of the indigenous raids.[29] Epidemics are reported as affecting indigenous populations in Chihuahua in 1545 and 1548 (Cramaussel, 1990a: 86), though the first recorded epidemic in the Nueva Vizcaya heartland per se was the 'peste' of 1577.[30] Such were the impacts of this catastrophic phase of disease that by 1579 the mines in Santa Barbara and Inde were described as 'almost depopulated'.[31] Another dramatic period of disease is recorded in 1590. By 1604, Bishop Mota y Escobar described the whole of Nueva Vizcaya as poorly populated and lacking in any substantial Indio towns, while Jesuit missionary and historian Andres Perez de Ribas, drawing on his experience of travelling in this northern region between 1604 and 1620, similarly highlighted the emptiness of the area (Gradie, 2000: 22–23). One must of course bear in mind that there had not been substantial Spanish colonization of the area by this stage and the indigenous population of this region was still predominantly semi-nomadic and scattered, thus perhaps giving the impression of a relatively 'empty' land anyway.

By taking away a source of labour and tribute, epidemic related indigenous depopulation in the sixteenth and early seventeenth centuries both necessitated and stimulated Spanish acquisition of land, particularly along the Conchos and San Pedro rivers (Deeds, 1989). The Santa Barbara area was a very fertile region, possessing plentiful water, timber, salt and saltpeter, as well as other resources which were used to support the mines (Griffen, 1979). Between 1563 and 1620, land in the area began to be distributed, allocated and settled (Cramaussel, 1990a). Some of the first residents of the area received awards of land to sow wheat and on which to grow fruits from about 1570 onwards, but it is clear that most of the first mercedes were awarded to soldiers who were encouraged to settle newly 'discovered' areas. Francisco de Ibarra conceded some land to his cousin, Martin Lopez de Ibarra, in April, 1570 (Rocha, 1942: 199), and to Santos Fernandez Rojo and Francisco Ruiz de Alarcón in May of the same year (Cramaussel, 1990b: 143). In accordance with merced policy, the land awarded had to be 'populated' to be validated, though this condition was clearly not always adhered to, and in 1598 the Governor of Nueva Vizcaya, Diego Fernández de Velasco, commissioned Captain Juan de Gordejuela to enforce landowners in the province of Santa Barbara to 'populate' newly awarded lands within one year of a grant being approved or risk their concession being made available for appropriation by others (Cramaussel, 1990b: 116).

Grants were also specifically awarded for water for irrigation, domestic uses, mineral processing, for running mills and for rearing livestock, specifically, in this area, cattle (Cramaussel, 1990a, 1990b; Alvarez, 1990). Access to water was, however, limited and had to be shared among a variety of users. All new requests for the use of water from a specified source had to be approved by existing users before a grant could be approved. This was frequently a protracted process, often entailing lengthy negotiations and sometimes resulting in bitter disputes.

Reports indicate that some haciendas had begun to emerge in the area by the close of the sixteenth century. These were for the most part 'Haciendas de labor' or 'Haciendas de pan llevar' to raise cereals, mainly maize and wheat, and were in the main restricted to the regions immediately surrounding the mining areas. In 1571, for example, there were thought to be seven haciendas, mostly producing maize, along the Florida River (Miranda, 1871), though a 1604 survey report on the Santa Barbara valley indicates that by that stage there were also ten cattle ranches and eleven grain farms in that area (Mota y Escobar, 1940, cited in Griffen, 1979: 44; Cramaussel 1990b: 123). By the turn of the seventeenth century, wheat, maize and various varieties of fruits and vegetables were being grown in conjunction with livestock grazing along the Rio Florido.[32] Concessions for keeping sheep had been granted in the San Bartolomé valley in the sixteenth century, but livestock ranches, mostly for cattle, were also found along the Conchos, San Pedro and Nazas rivers. The estancia belonging to Pedro Sánchez de Chávez, which consisted of five sitios for cattle and four caballerías of land, was among the most important in the region around this time (Deeds, 1989).

The amount of land under cultivation increased throughout the seventeenth century and the production of maize and wheat increased from 6,300 fanegas[33] in 1604[34] to over 40,000 fanegas by mid-century.[35] Yet there was no equivalent increase in the number of agricultural properties, largely because

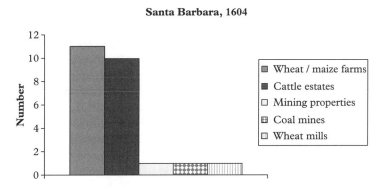

Figure 3.1   Land use and property in Santa Barbara, Chihuahua in 1604 (Urdinola's 1604 survey: Borah, 1955)

of the limitations imposed by water availability in this region. Rather, there was a process of consolidation. In the Valley de San Bartolomé, there were 15 estancias in 1622, 13 in 1632 and only 12 by 1654. Indeed, between 1620 and 1700, there was a gradual process of accumulation of territory by individuals and the various branches of the Church (Cramaussel, 1990b: 122). There was further consolidation between 1700 and 1760 (Cramaussel, 1990b; Alvarez, 1990) and the most fertile and best watered lands began to be monopolized by a relatively small number of people. Deeds (1989) considers one case, that of Sargento Mayor Valerio Cortés del Rey. Through

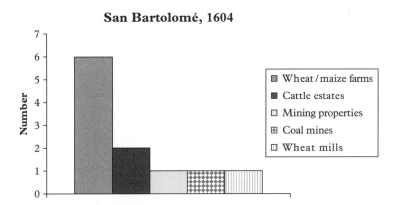

Figure 3.2   Land use and property in San Bartolomé, Chihuahua in 1604 (Urdinola's 1604 survey: Borah, 1955)

accumulation of merced grants, together with purchases of over 25,000 ha along the Conchos River, Cortés del Rey established one of the most important beef ranching haciendas in the region and one of the most significant properties in the whole of Nueva Vizcaya. He also developed his mining interests in the Valle de San Bartolomé, increasing his mineral production from 3,000 cargas to over 40,000 in 1640. By 1679, his estate was thought to be worth around 300,000 pesos (Aboites Aguilar, 1994) and his properties supplied meat, milk and grains for many of the mining towns and would later become part of the Cortés del Rey Mayorazgo in 1689.[36]

'Pestes' of what are now thought to have been more widespread epidemics of measles and smallpox continued to take their toll on indigenous populations throughout the seventeenth and eighteenth centuries and particularly on the younger members of society. Epidemics are reported in 1610, when a third of the population of the province of Santa Barbara is thought to have been lost (Rodriguez, 1987: 190), between 1616 and 1620, coinciding with a major Tepehuan revolt which will be considered in chapter 5 (Gerhard, 1982: 169; Gradie, 2000: 24), between 1662 and 1663 (Orozco y Berra, 1938), in 1666 and again in 1676.[37] Notwithstanding these epidemics, the population grew as a result of an influx of people attracted by the mineral prospects in the region. Silver strikes at San Diego de Minas Nuevas to the west, for example, and also in Santa Eulalia just to the southeast of the modern-day city of Chihuahua in the second half of the seventeenth century, led to an expansion in mining activities and stimulated the founding of the city of Chihuahua in 1709 (renamed San Felipe el Real de Chihuahua after 1723). By 1716, there were 1,700 inhabitants in Santa Eulalia and Chihuahua, by 1725, the figure was 2,500 and over 20,000 inhabitants were living in the two settlements by the middle of the eighteenth century (Cramaussel, 1990b: 121). There were an estimated 80,600 people living in Chihuahua by 1800 (Gerhard, 1982), though it should be noted that all available population estimates are imprecise and tend to under-report the groups of lower social status (Martin, 1996: 26).

By 1750, the population of northern Mexico as a whole was estimated to be around 350,000, of whom approximately 64 per cent were indigenous. The population of Nueva Vizcaya, with 124,000, however, was the most populated province (Aboites Aguilar, 1995: 35). By 1763 there were an estimated 202 different families in the San Bartolomé valley alone, amounting to some 1,833 persons. Some 89 Indio families lived in the mission town of San Francisco de Conchos, which was home to 289 non-indigenous residents (Tamarón y Romeral, 1937).

Throughout the eighteenth century there was a steady expansion in agricultural exploitation and stock raising to support the growing population

(Cramaussel, 1990b: 121), with a concomitant increase in the area under irrigation. The best land and water sources continued to be monopolized by a limited number of landholders. Two large estates are highlighted in the documentation of the first half of the eighteenth century around the Rio Conchos area: Domingo del Valle, which comprised 45 individual sitios, and the slightly smaller Francisco Duro, which was made up of 35 sitios. Both were managed by grain merchants from San Felipe el Real de Chihuahua (Alvarez, 1990).

Benefiting from generous 'donations',[38] the Jesuits also became important landholders in the north of Mexico (Deeds, 2003). Details of the redistribution of Jesuit properties, following their expulsion from Mexico in 1767, reveals the extent of their monopoly up to that point. One document from 1773, for instance, deals with the Hacienda de Nuestra Señora de los Dolores in Chihuahua, which according to merced titles, comprised 20 and a half sitios for cattle grazing, along with 21 and three quarter caballerías. Each of the sitios was estimated to be worth some 250 pesos and to consist of good quality land for grazing and pasturage, all in all worth around 5,000 pesos. Each of the caballerías of land was valued at 1,100 pesos and the land included access to 'abundant water' sources, estimated to be worth in the region of 23,375 pesos. The lands also incorporated 'a spring with its irrigation ditch and other channels', valued at approximately 3,000 pesos. The whole of the hacienda with its sitios, caballerías and resources was at the time thought to be worth 31,375 pesos.[39] Much of the land acquisition in this region contributed to the formation of smaller though still substantial landholdings of between 800 to 4,000 ha. Nonetheless, by creating a highly unequal distribution of land and resources, this feature of the colonial political economy effectively exacerbated the differential social vulnerability of society in the region to periods of climatically induced hardship and crisis (Liverman, 2000), and particularly prolonged drought.

While population expanded generally throughout the eighteenth century, there was a steady decline in the indigenous populations of the region (figure 3.3). By this stage many of the indigenous groups had been fairly well assimilated and acculturated into the colonial enterprise. Indigenous groups had thus become more of a minority in this region and were increasingly restricted to mission towns and small areas of Spanish settlement (Griffen, 1979). However, there was significant hostility between some local groups and the Spanish in the region throughout the colonial period. Drought, food scarcity and epidemic disease may have played a significant role in triggering or at least contributing to some instances of indigenous unrest and revolt, an issue which will be discussed in chapter 5.

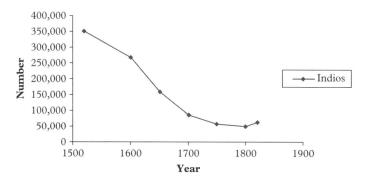

Figure 3.3   Indigenous population, Nueva Vizcaya, 1521–1821 (data included in Gerhard, 1982)

From frontier to breadbasket: creating a colonial society in the Bajío

Like the north of Mexico, the main stimulus to colonize the Bajío came with the discovery of mineral (silver) wealth, firstly as shown north of the region in Zacatecas, and then later in the mountains of Guanajuato. The early opening up of highways from Mexico City and Michoacán to the mining areas further north, plus the use of wagon trains on these routes, made imperative the establishment of safe stopping points for merchants, caravans and traders heading to and from the mines. A series of garrison towns was therefore initially established to safeguard the first settlers, and also the passage of mineral wealth through the area, from Chichimec attacks.

Four forts were initially established in Guanajuato.[40] San Miguel el Grande in Guanajuato, however, was the first presidio town to be established, in 1555. This was a former indigenous settlement (also housing some Spanish settlers) that had been abandoned due to Chichimec attacks by the time Viceroy Don Luis de Velasco ordered the presidio to be founded.[41] The site was located in the San Miguel Llanos (or 'flatlands'), at the 'crossroads of Chichimeca raiding forays' (Powell, 1944: 183), and was settled to guard against Chichimeca raids on caravans on the road from Mexico City to Zacatecas and to provide 'security for the said road'.[42] Velasco commissioned Angle de Villafane to found the town on his behalf,[43] and to populate it with 50 Spaniards who would be allocated lands for houses, orchards, farms and livestock on territory that was set apart from the peaceful Purépecha, Chichimeca and Otomi communities who had already settled nearby.[44] Indios from the nearby towns of Huango, Cuitzeo, Acambaro and Querétaro were drafted as labour to build the town.[45] As an incentive, each new resident received a grant of land consisting of two caballerías (about 33 acres),[46] but it was a requirement for the colonists (including Indios) to arm themselves against the inevitable Chichimec attacks before their grant of land was approved.[47] In 1562, another defence town, San Felipe, was established, possibly as a direct response to a large Chichimeca uprising the previous year (Powell, 1944: 184). Similar defensive towns were founded elsewhere, including Léon, latterly the capital of Guanajuato, in 1576 (Brading, 1978: 16). Notwithstanding the establishment of such presidio settlements, however, attacks and raids continued (Moreno, 1958: 85–87).

As with the north of the country, missionary activities were central to the pacification and colonization process in the Bajío. Franciscans accompanied the first Spanish conquistadores and colonists to the region, founding convents from the 1530s, but the evangelization of the region really began in earnest after 1565. The convent at San Francisco was established in 1573, though amid further attacks – two priests from there, Doncel and Burgos, were murdered by Chichimecas while travelling around San Felipe (Moreno, 1958: 88–89) – and by 1585, convents had been established at Querétaro, Celaya, San Felipe, Tuliman and Apaseo,[48] though the missionaries continued to live with the threat of violent Chichimec raids. Léon was almost depopulated as a result of attacks in 1589 and the missionaries, fearing for their lives, fled. Nonetheless, this town more than many of the others appears to have been able to withstand these disturbances and flourished from its initial population base of 50 people, to become an Alcadia Mayor only three years later.[49] In the same year as these attacks, the Jesuits, heading north from Pátzcuaro, embarked upon a mission to reach the uplands of San Luis Potosi, to the modern-day municipality of Dolores Hidalgo, areas in the vicinity and also around Guanajuato (Moreno, 1958: 91). Thereafter, Spanish presence, and influence, in the area gained strength and the number of attacks decreased.

Evolution of the agrarian heartland

In the early years of the colony, a number of encomiendas were awarded in the Bajío, mainly in the more fertile southern areas of Apaseo, Yuriria and Acambaro. Juan de Villaseñor received an encomienda in Huango which consisted of four plots, measuring close to 3 leagues in total, while Hernán Pérez de Bocanegra, a powerful and well-connected Spaniard, received encomiendas in Apaseo and Acambaro in 1538. The latter was in effect an Otomi pueblo, adjacent to Acambaro and an attractive site, having an important natural spring that emerged close to the Rio Querétaro (Murphy, 1986; Wright Carr, 1998: 39). The first actual land grants awarded in the Bajío, however, date from 1542,[50] and a number of mercedes for estancias, for cattle and sheep began to be introduced to the region during the 1550s.[51] Individuals like Bocanegra would benefit from the accumulation of mercedes in such areas,[52] but early mercedes records reveal that some local Otomi caciques, such as Diego de Tapia, son of the founder of Querétaro, also became particularly wealthy landowners in the area, with investments in wheat production, sheep farms and mining (Brading, 1978: 16).

The discovery of silver in the sierras of Guanajuato, coupled with indigenous depopulation, the loss of both hostile and tribute paying Indios and the consequent availability of land, stimulated greater agricultural investment in the area and the number of both agricultural and livestock awards began to increase in the 1560s and 1570s. There is very little information available with which to trace the actual rate and scale of indigenous depopulation in the Bajío, not least because much of the region was occupied by largely nomadic groups. There were a number of indigenous settlements in the south of the region, inhabited predominantly by Tarascos and Guamares (one of the Chichimeca groups). Towns like Acambaro, Yuriria, Huango, Puruandiro and Apaseo faced dramatic population decline. In Yuriria, for example, estimates around the middle of the sixteenth century suggest a total of around 9,000 tribute paying Indios. By the time the Relación was compiled, however, there were only an estimated 900 tributarios, a decline of between 80 and 90 per cent (Boissonas, 1990). This decline led to land abandonment and more land became open to appropriation.

The first land grants awarded in the vicinity of Yuriria date from 1550, but by the close of the sixteenth century totalled 81, covering an area of some 114,000 hectares for livestock grazing and 5,461 hectares for agriculture (Boissonas, 1990: 48). Some estancias were awarded in conjunction with small plots of agricultural land,[53] though mercedes intended solely for agricultural land were also approved. There was some trading and purchase of mercedes, despite conditions stipulating that property was not to be sold, or transferred after the award was made. There was a particularly high turnover of land in the Yuriria-Salamanca area.[54] Merced 'donations' also allowed certain religious sectors to gain a foothold on the property market. In 1643, for example, Don Diego Delgado donated a substantial property to the Augustinian convent in Salamanca (see below).[55]

A number of towns were established in part to assist in the agricultural development of the region. Celaya was founded in 1571 at the confluence of the San Miguel and Apaseo rivers. This was clearly a good defensive position and the presence of the town could safeguard routes of mineral transportation to and from Zacatecas and Guadalajara.[56] But Celaya was primarily established to encourage agricultural development of the fertile land alongside the Rio Lerma. As had been the case with other fledgling settlements in the region, each new settler was granted 2 caballerías each, lots for a huerta or garden, a vineyard and a plot upon which to build a home, though on condition that they remained resident in the town for a ten-year period with no absences longer than four months (upon penalty of losing his/her lands). At least thirty married Spanish residents were attracted by these incentives (Rodrigues, 2005). Once again, labour was drafted from the nearby towns of Yuririapundaro, Acambaro, Ucareo, Zinapecuaro and Cuitzeo to assist with the construction of the urban infrastructure,[57] and it is clear that the town prospered and grew rapidly. The Relación Geográfica for Celaya, for example, records an increase from three or four residents in its first year to sixty by the time the Relación survey was conducted in 1579/80.[58]

In 1602, the new town of Salamanca, between Celaya and Léon, was similarly founded to function solely as an agricultural centre, with more generous awards of 4 caballerías awarded to its new settlers. Spanish colonization of towns like Celaya and Salamanca and the awarding of agricultural land grants in the area attracted further substantial migration of Purépecha and Otomi Indios from the south and east, some of whom were awarded lands of their own. Others were welcomed by the estate owners as tenants and labourers (Brading, 1978: 17). Not all migrants relocated voluntarily. Repartimientos of forced Indio labour from pueblos in Michoacán were drafted to some of the first residents of these new towns to assist in the building of houses and to provide labour for public works programmes.[59] This practice appears to have continued throughout the second half of the sixteenth century,[60] judging by claims forwarded by Indio communities at the turn of the seventeenth century that this was leaving their own original pueblos 'almost depopulated'.[61]

There were variations in the type of land uses that developed in different parts of the region. By 1591 it is estimated that 153 awards of agricultural land had been granted in the vicinity of Celaya (Butzer and Butzer, 1993), while by the same year, an estimated 41 estancias had been formally approved, mainly for cattle estancias, in the lands between Celaya and Léon. Thirty-two cattle ranches had also been granted in the vicinity of Guanajuato by this time and sheep were also raised in the region, specifically around San Miguel. After 1591, and reflecting the Crown's livestock marginalization policies of the period, however, cattle would be marginalized to more western regions, though in greatly reduced numbers, while sheep transhumance expanded northward (Butzer and Butzer, 1997: 170). Nonetheless, around the turn of the seventeenth century, there were a number of important livestock ranches in the region. Diego Frausto, a resident of Léon, claimed to have 8,000 cattle in his possession in 1606,[62] while Bernadino Guerra of Celaya had 7,000 sheep and Antonio de Vivieca 16,000 in his estancia known as Calera in Apaseo (Lopez-Lara, 1973: 157, 162).

It was for agricultural production, however, and specifically the cultivation of irrigated cereals, for which the Bajío gained notoriety. Although there had been some cultivation by Otomis and Tarsacos, it was really only after Spanish colonization that agriculture expanded across the region. Cereals were cultivated around Celaya, where in 1580 between 17,000 and 18,000 fanegas of maize had been produced. By 1600 the figure was 30,000, and by the middle of the seventeenth century production stood at 150,000 (Chevalier, 1952), representing a 70 per cent increase between 1580 and 1600 and a very impressive 400 per cent increase from 1600 to 1650 (Boissonas, 1990: 131). Awards of mercedes for both agricultural land and for livestock ranches decreased generally after the 1630s (Brading, 1978: 17), and although mercedes were still being awarded, they were much more commonly granted for specific associated purposes, notably to secure access to water for milling and irrigation purposes.[63]

The awards that were made in the period between the 1550s and the 1630s were influential in determining the nature and form of agrarian development of the region. Colonial elites accumulated wealth through commerce and mining in this region and, in effect, secured this wealth through the development of vast and valuable landed estates (Tutino, 1998: 373). In some cases this process of accumulation can be traced over several hundred years. The estancia of Juan de Aranda in Irapuato represents one such example. Although Aranda's property would in fact change hands, the name of the estate would remain the same, so allowing us to trace the process of territorial accumulation. An estancia was first awarded to Juan de Aranda during the period of office of Don Luis de Velasco. Juan and Pedro Aranda sold the property (including houses, stock and equipment) in a consortium with Maria Villar, wife of Pedro, to one Don Pedro Lorenzo de Castilla in 1578. Castilla, whom Brading (1978) considered to be one of the wealthiest men in Guanajuato at the time, increased the size and value of the property through additional purchases. The estate was still intact by 1792, when there were 35 resident landlords, 67 tenants and a total of 112 people living there (Martinez de la Rosa, 1965: 64–66). The last wills and testaments of other individuals similarly reveal the scale of property ownership in the vicinity of Léon. The will of Don Diego Antonio de Quijas y Escalante, a local resident, for example, describes his estate as comprising 2 estancias for cattle grazing, 2 for sheep, 65 caballerías, houses, wells and water wheels. Upon his death in 1787, Escalante's estate was thought to be worth in excess of 26,901 pesos.[64]

The rapid expansion in the exploitation of land in the Bajío, however, and the emergence of landed monopolies, was very much dependent on adequate supplies of water. The pattern of heavy but variable summer precipitation, often preceded by hot, dry conditions in spring, influenced the type of agriculture that developed in the region. Maize flourished in this climate whereas the preferred cereal crop, wheat, which the Spanish introduced, required irrigation to survive the dry winter months, while livestock raising was of course also obviously dependent upon a regular supply of water. For this reason, many of the first land grants for livestock or agriculture were awarded alongside rivers, streams and springs to ensure water access.[65] Permanent and ephemeral watercourses, rivers and arroyos all began to be exploited and it is clear that groundwater was also tapped in many locations, though perhaps more so in the later colonial period.[66] There was extensive use of water storage, diversion and water management systems. Storage of floodwater, for example, for use during the dry season, a strategy referred to in the documentation as medio riego, is thought to have been an early practice.[67]

Examples of direct acquisition of water rights are few in number and the amount of water in question was usually limited (Murphy, 1986). In Celaya in the 1570s, however, nine of the twelve grants awarded to residents of the new town included a right to access water on certain days. Seventy one caballerías of land with water for irrigation were awarded in the vicinity of Yuririapundaro between 1583 and 1590,[68] while of the 57 mercedes awarded in Salamanca between 1608 and 1635, 31 included water rights for irrigation purposes (Boissonas, 1990: 65). Provision of water for irrigation was of course also a central requirement in the establishment of some of the new settlements in Guanajuato. Salamanca was founded on the basis that the residents would construct a water channel for irrigation,[69] although because of a lack of capital, the channel had still not been built by 1619 and extra labour had to be drafted in to complete the work.[70] In some places, water sharing arrangements were drawn up early on. In Celaya, a fairly egalitarian system of water distribution had been established by 1575. Water rights were allocated by shifts, or tandas (Lipsett-Rivera, 1999b: 20). The Viceroy suggested that water should be distributed 'equally by shares or days … among the said fields so that one person will not receive more than another, bearing in mind the need to reserve enough water for future settlers.'[71] Under this theoretically sustainable scheme, each resident user was granted two days of water (Murphy, 1986: 18), a policy which remained in place into the early eighteenth century.[72]

The archives reveal that indigenous communities were quite successful in securing water rights. In 1591, the community of Yuriria had constructed water channels, ditches and ducts to conduct irrigation water to their fields.[73] The people of Acambaro filed a request to drain water from the Rio Grande for their fields in the same year,[74] and newly founded Indio settlements across the Bajío, such as San Francisco del Rincon (Nuestra Señora de la Concepción del Rincon), established in 1651, were also provided with water for irrigation (Boissonas, 1990: 134). Mercedes were also awarded for remanientes, literally 'left over water',[75] such as that granted for landowners around the township of Santiago Tarandaquaro (in the vicinity of Celaya) in 1614,[76] though the concept of remanientes was unavoidably vague and sometimes resulted in disputes. One such case involves the sharing of rem-anientes between the indigenous community of the town of Acambaro and Francisco de Villadiego Senderos, beneficiary of Indaparapeo. According to Senderos, the Indios of Acambaro had attempted to remove all the leftover water from the river of Tarandaquaro for the running of a mill to which Senderos held water rights according to an official award of a merced. In this case, an agreement was drawn up between the two parties to share the use of the water. The water was to first go to run Senderos' mill and the used water would then be channelled to irrigate the community's fields.[77]

The scale of water management in the region expanded significantly in the eighteenth century (Murphy, 1986: 33). Many earthen and brick built dams, aqueducts, canals and reservoirs and other water control devices were constructed and water was diverted from key rivers in the region such as the Rio Laja and the Lerma to irrigate wheat fields or to provide water for towns (Murphy, 1986). As Murphy (1986) has already demonstrated, somewhat surprisingly, colonial Mexico was only partially connected with European centres of scientific innovation and, as a result, much of this water technology was constructed by people with little or no training. As shall be discussed in chapters 4 and 5, the results were consequently somewhat mixed. There were regular dam breaches and problems with maintenance of acequias or drainage ditches, leading to many cases of inundation, some of which in turn stimulated bitter and often protracted lawsuits between landowners.

Population expanded rapidly in the Bajío, but especially in the eighteenth century. The very rich sources of silver and gold at Santa Fe would encourage an expansion in the number of buildings, the urban infrastructure and also in an increase in population. In 1619 Guanajuato was referred to as a town, and by 8 December 1741 it had officially gained city status. The most important mines in the district employed 10,000 people by the middle of the eighteenth century (Humboldt, 1811: 161–162) and by 1741 the city itself is thought to have had a population of around 40,000,[78] although the intendencia of Guanajuato (an area that approximates with the modern-day state) supported 156,140 people by 1742. By 1793 the figure had risen to 397,942 and there are estimates of around 513,300 just after the turn of the nineteenth century (Humboldt, 1811). By 1803, Guanajuato was the fourth largest province after Mexico, Puebla and Oaxaca, but Guanajuato City was the third largest city (after Mexico City and Puebla) with 32,098 inhabitants (Humboldt, 1811). According to data from about 1792, only about 44 per cent of the population in the Bajío was indigenous, though figures vary across the region (figure 3.4). In the mining zone of Guanajuato, as might perhaps be expected given the control they held over mining operations, there were many more Spaniards than other sectors of the population. Moreover, population growth was not uniform across the region as a whole.

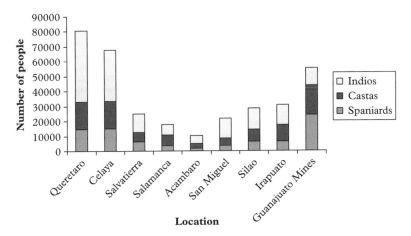

Figure 3.4   Composition of population in the Guanajuato Intendancy, c. 1792 (after Tutino, 1986: 75)

Between 1760 and 1810 the population in the Bajío generally increased by 33 per cent.

This general population expansion and the availability of land in the region, coupled with continued commercial mining booms, resulted in more land being brought under cultivation in the late seventeenth century and throughout the eighteenth century. Wheat, fruits and vegetables became increasingly profitable (Tutino, 1998: 376). This would transform the nature of land use in the region and the economy shifted further away from livestock ranching to agricultural production. In fact, by the middle of the 1780s, a substantial proportion of the population (48.6 per cent) was employed in the agricultural sector.[79] The value of land rose appreciably and the production of cereals, specifically maize, increased across the region, accounting for approximately 45 per cent of total tithe value by the late eighteenth century (Rabel Romero, 1975). Many of the estates in the region expanded their maize production, often bringing more marginal land into use for the purpose. This would ultimately result in declining maize yields as these more marginal tracts of territory lost their fertility (Tutino, 1998: 376). Guanajuato, and in fact the Bajío region as a whole, became relatively prosperous based on agricultural production, mining, industry, trade and to a lesser extent, stock raising. Such was the rapid rise of economic development in the Bajío that the region has been referred to as the 'pacemaker of the colonial Mexican economy' (Brading, 1978: 18), and as Tutino (1998: 375) has suggested, 'no region became more pivotal to the colonial economy in Mexico, especially in the eighteenth century, than the Bajío'.

Spaniard meets Indio: Spanish colonization in the Valley of Oaxaca

The first Spanish settlement in Oaxaca was established on land northeast of the garrison of Huaxyacac (Gay, 1950: 1, 390). This town became the chief residence for Indios who participated in many aspects of urban life and business, while other indigenous satellite towns were mainly agricultural (Chance, 1976: 611). The Spanish community in the valley gained a stronger position after 1529, when the First Audiencia appointed Juan Pelaez de Berrio as Alcade Mayor of the province of Oaxaca. Berrio proposed a plan to develop the Spanish presence in the valley which included the demolition of Huaxyacac and the building of a new town, Antequera (modern-day Oaxaca City), on the same site. Antequera came to resemble other Spanish towns, typified by the traza (grid iron street plan) and surrounded by settlements housing indigenous communities. As was true of the establishment of new settlements in the Bajío and elsewhere, the town was constructed using indigenous labour and the relatively small Spanish population relied on the food and commodities supplied by tribute paying communities. The local economy was based upon agriculture, stock raising and trade, but the city represented an important political, religious and administrative centre (Chance, 1976).

The Valley of Oaxaca was generally spared the problems associated with encomiendas. In 1529, however, Hernando Cortés received the title Marquesado del Valle. This was a royal grant encompassing much of the land of the valley. Commonly referred to as the Marques de Quatro Villas, he held quasi-feudal control over the land and people of Etla, Oaxaca, Cuilapan and Santa Ana Tlapacoyan (Bustamante, 1992). Like other encomenderos, the Marques could demand the tribute and labour of the indigenous populations within his jurisdiction, although he could not legally expropriate lands already under cultivation. Moreover, his own interests in the region were limited to exaction of tribute rather than forced labour. He had the additional privilege of being able to distribute lands as encomiendas to others,[80] though he chose not to, not least because this strategy secured his monopoly over territory in the area. There was a gradual reduction in the size of the territory subsumed under the Marquesado by the middle of the sixteenth century as a result of the consolidation of royal authority in Mexico and difficulties between Cortés and his rivals (Taylor, 1972: 36). By this stage, changes in the administrative system had started to challenge the exploitative behaviour of many encomenderos. Nevertheless, a number of other encomiendas did still exist in Oaxaca in the sixteenth century and as community complaints suggest, the encomenderos may have continued to exploit their position. The community of the town of Jutla, for example, had been required to provide their services in the mines of Chichicapa near

Antequera, the number of individuals expected to participate in repartimiento each week having been fixed at 4 out of 100 people, that is to say, 4 per cent. By 1603, however, the community's population had been dramatically affected by the epidemics that would sweep across the country in the second half of the sixteenth century. A request from the community to the Alcades and Regidores of Oaxaca notes that 50 people had died and that the repartimiento demands should be modified accordingly. The solution the community forwarded for consideration was that the demands stay at 4 per cent, but out of the remaining population.

Beyond such indirect references, actual data to trace the extent of indigenous depopulation in Oaxaca following European contact are fragmentary. Most of the indigenous towns declined in size between 1520 and 1630 (Taylor, 1972), but the population of the valley may have fallen from an estimated 350,000 on the eve of Spanish conquest to about 150,000 by 1568. Total population may have dwindled to an all time low of around 40,000–50,000 by the third decade of the seventeenth century (Taylor, 1972). A number of existing towns disappeared altogether during the sixteenth century (Taylor, 1972: 27), while others were reconstituted or relocated. In 1591, for example, the population of the town of Chalcatongo, which at the time fell under the encomienda of Carlos de Luna y Arellano, petitioned to relocate to a healthier place. The community had suffered dramatic population losses, 'most of the people had died and those that were left were ill and each day there were more dying'. Their solution was to relocate to 'Santa Cruz Xocoyolaticalpa' which was regarded as healthier, 'where there were good lands, waters and trees, a good church' and which, they added, fell under the jurisdiction of the same encomendero.[81]

Dominicans arrived in the Valley of Oaxaca in 1528 and the first Bishopric of Oaxaca was founded at Antequera only seven years later. Supported by the Crown and the Marques del Valle, doctrinas, religious head towns and subject communities or visitas, had been established in Etla, Cuilapan and the Villa de Oaxaca by 1550, in Huitzo in 1554 and Ocotlán in 1562. By the close of the sixteenth century there were over seventy Dominicans in the valley, although Franciscans, Jesuit, Augustinian, Carmelite and Mercedarian orders were also present in Antequera. In order to facilitate evangelism and administration of the Indio populations, the Dominicans instituted a network of congregaciones. There are thought to have been nine civil congregaciones in the Valley of Oaxaca around the turn of the seventeenth century, some with repartimiento grants attached to them. The indigenous caciques may have been pivotal to the consolidation of Indio towns in the early sixteenth century, taking a leading role in the congregación of indigenous subjects (Taylor, 1972), although most of the congregaciones had disaggregated into their constituent communities by the close of the seventeenth century.

European and indigenous land use in colonial Oaxaca

Mercedes for grazing or agriculture were not awarded in Oaxaca until 1538, and, to begin with, Spanish land use appears to have been exclusively for the raising of livestock (Taylor, 1978: 71–105).[82] The region was valued for transient livestock grazing and there is evidence to indicate that the Spanish mesta system of 'sheep walks' or long distance seasonal treks had been introduced to the region as early as 1542.[83] By the 1550s the extensive tierras de humedad or marshy lands alongside rivers such as the Atoyac appear to have been particularly valued for seasonal grazing,[84] and many unused grassy areas were gradually turned over to more permanent grazing for cattle, horses and sheep, goats and pigs. These sites were often adjacent to agricultural plots, located so as to make the most of fertile alluvial soils and water for irrigation. It is not surprising, therefore, that from the middle of the century there were many complaints of crop depredation as a result of livestock straying from estancias.[85] Nonetheless, large sheep and cattle stations were in operation in the valleys by the 1560s, with cattle concentrated in the Zimatlán region and sheep in Talcolula.

Beans and maize continued to be grown in the Valleys of Etla and Tlacolula after conquest. Judging by the number of mercedes for wheat mills that were granted in the first hundred years of colonial rule,[86] however, wheat appears to have been rapidly adopted, and by indigenous communities as well as Europeans.[87] The Valley of Etla with its permanent supply of water from the River Atoyac and fertile alluvial soils provided ideal conditions for irrigated wheat and this area had become the main cereal producing area of Oaxaca by the late sixteenth century.[88] The chief European products from the region, however, also included lentils, lettuce, cauliflower, onions, garlic, radishes and various fruits.[89]

Sugar cane was grown, specifically in Jesuit landholdings in the Valley of Tlacolula, though with some difficulty against a backdrop of climatic variability and unreliable rainfall.[90] Silk from Oaxaca was traded internationally, and despite a general downturn in the silk trade in the 1600s (Borah, 1951),[91] it remained an important commodity at the end of the eighteenth century.[92] There is debate as to whether salt was produced in pre-Hispanic Oaxaca (Hewitt et al., 1987; Doolittle, 1989). References to salt production in the sixteenth century, however, do appear in the Relaciones Geográficas for Mitla and Tlacolula, and after the conquest, salt deposits (salinas) across the region began to be exploited commercially, though as shall be illustrated in chapter 4, flooding would affect this sector of the economy on a number of occasions.[93]

One of the most important commodities produced in colonial Oaxaca was cochineal, the red dye that is derived from the dry bodies of the Dactylopius coccus, an insect indigenous to southern Mexico. Cochineal was traded in Tenochtitlán, was among the tribute commodities that

Aztec rulers exacted from conquered subjects and was used as a dye for fabric, stone and wood (Baskes, 2005). The Relaciones Geográficas from the sixteenth century refer to the properties of the 'grana cochinilla' (coccus cacti) as being known in the indigenous pueblos from 'time immemorial', and not only in Oaxaca but also in Chiapas, Guerrero, Puebla and Tlaxcala (Viruell, 2004). Soon after European contact, measures were implemented to develop its production in these areas, but it was in Oaxaca that the insect thrived and production in Oaxaca soon displaced that of the other regions.

The prevalence of the nopal cactus favoured by the cochineal insect, particularly in the southern arm of the valley, where two harvests could be secured each year, ensured an alternative income, and at times, a basic livelihood, for those with only limited access to good land and water supplies (Taylor, 1972: 47, 94). Cochineal production did not require especially good farmland, could be produced in the smallest of backyard cactus groves, and is thought to have helped offset problems of land shortage for communities in the southern arm of the valley. In 1609, for example, it was reported that the town of Ocelotepeque in the jurisdiction of Miahuatlán had poor quality land and 'very little maize is sown'. Scrub was burned, as was common practice to improve the land quality, but the only successful product that could be 'raised' was cochineal.[94]

The production of cochineal was unusually dominated by indigenous populations. Although there were some failed attempts by Spanish hacendados to enter into cochineal production (Donkin, 1977),[95] it was only local indigenous populations who 'had the patience and dedication ... and unique capacity' to produce cochineal dye.[96] This level of what Blaikie et al. (1994: 63) might term 'ethnosience' was perceived to be critical to the success of the cochineal industry. The insects could easily fall prey to birds and other insect predators and so constant vigilance over the cacti and the insect populations was essential (Baskes, 2005). The production of the dye was an extremely labour intensive and painstaking process and the cochineal produced in Oaxaca was thus classified as 'grana fina', reflecting the time and effort invested in its production.

Cochineal began to be purchased by Spanish merchants to ship back to Europe as early as 1540 (Baskes 2005: 11). Production increased towards the close of the sixteenth century, but was severely disrupted by indigenous population decline between 1590 and 1650 (Pastor, 1987; Frizzi, 1991). To counteract the decline, the Crown instituted a production scheme based on credit that encouraged cochineal production among the remaining indigenous communities. Where the local populations dominated the production of the dyestuff, the Spanish played central roles as financiers and exporters. Local Spanish district magistrates and administrators, normally the Alcades Mayores, provided credit for cochineal producers against a future delivery of the product in the form of 'repartimientos de bienes', that is to say, goods advanced on credit and repayable by either coin or kind.[97] This scheme, together with indigenous repopulation, facilitated an increase in cochineal production between 1660 and 1700 (Viruel, 2004). By the eighteenth century, the cochineal industry of Oaxaca had become an important economic mainstay of the region and after silver was Mexico's most valuable export commodity.[98] There was something of a boom in the industry after 1740 (Taylor, 1972: 183), as demand in Europe grew,[99] and particularly in the years leading up to 1787, when around 30,000 arrobas[100] were produced annually.[101]

As Baskes (1996, 2005) has demonstrated, there is a considerable literature that focuses on the exploitative nature of this trading system between Indio and Spaniard. However, new interpretations of the system suggest it might have provided an important risk avoidance strategy for many indigenous communities (Ouweneel, 1996; Pietschmann, 1988; Baskes, 2005). Although in some situations creditors could seek jail sentences for 'non-payment' of goods, or could seize property and possessions from debtors in lieu, it seems that the repartimiento system in Oaxaca at least might have actually worked in the interests of the indigenous population. Loans made by Alcades Mayores for cochineal production were interest free and made an important contribution to the indigenous rural economy.[102] The system prevented the Indio cochineal producer being subject to fluctuations in market demand and smoothed out consumption during the agricultural cycle.[103] In this sense, all the risk was carried by the creditors and only marginally so by the indigenous communities responsible for cochineal production (Baskes, 2005).

Output of cochineal, however, could be highly variable. The extreme vulnerability of cochineal to climatic variability and inclement weather, such as an unexpected frost, excessive wind, rain or drought, had the potential to destroy entire harvests, so making repayment of loans and credit impossible in some situations (Donkin, 1977; Baskes, 2005), though some producers might have capitalized on this recognized vulnerability in order to avoid delivery of cochineal for which they had received advance payment (Baskes, 2005: 196).[104] As shall be illustrated in chapter 6, this vulnerability together with various other factors, including the abolition of the repartimiento trading relationship as part of the Bourbon reforms introduced by Viceroy Galvez, combined to cause a dramatic decline in the cochineal trade by the early nineteenth century. This would have implications for the social and economic well-being and resilience of indigenous populations of the region as well as the former financiers.

Indigenous land retention in the Valley of Oaxaca

Records from Oaxaca indicate that the Spanish or Creoles[105] increased their land investments after 1630, and amalgamated land grants into large estates and ranches (Taylor, 1978: 77), many of which were clustered around the most fertile and populous regions of the valley (Hamnett, 1971). The Church too began to accumulate territory. Once they had consolidated their stronghold in Oaxaca, through donation, capellanías and gifts from local communities and individuals, the Dominicans acquired many properties throughout Oaxaca. Not all of the native donations were necessarily made willingly, but as a result of persistent coercion by the Dominicans (Taylor, 1972: 169), with the caciques often mediating in such situations. In 1644, for example, the community of the Villa de Etla raised concerns that most of their lands had been so 'donated' to the local Dominican monastery by caciques and principales.[106]

By the end of the sixteenth century the Dominican monastery at Antequera was one of the main landowners.[107] By 1713, the Dominicans owned 33 sitios for grazing and 21 caballerías of agricultural land around the Bishopric of Oaxaca,[108] and the extent of their holdings continued to grow until the end of the colonial period. The monastery at Cuilapan, founded in 1555, held six sheep ranches, a farm and property in Antequera at the end of the sixteenth century, though had acquired more by 1618. Most of the land, however, was rented out (Taylor, 1972).

Pressure towards increasing secularization would eventually reduce the power of the Dominican church in the later part of the eighteenth century. By 1760, the Dominicans no longer held Etla, Zimatlan, Tlalixtac, Zaachila and Cuilapan, which were thought to be the richest of the valley parishes.[109] Although the Franciscans held a house in Antequera, their vow of poverty and non-investment in the land meant that they owned no rural or urban property in Oaxaca by the end of the eighteenth century. The Jesuits, who had founded a college at Antequera, held little in the way of rural investments, apart from a small wheat farm to the southeast of the city and several rural estates on the outskirts of the valley, though they did invest more substantially in urban property (Taylor, 1972). The Carmelites owned some holdings in the valley, including the hacienda of San Juan Bautista and the neighbouring estates of Los Naranjos and San José, which was valued at 34,240 pesos when the hacienda was sold to Juan Martinez de Antelo in 1745 (Taylor, 1972), while other religious orders, such as the Bethlemites, and various nunneries, had also become relatively prosperous landowners in Oaxaca by the late colonial period.

Generally, however, the Church in Oaxaca did not take an active interest in investing in agricultural land.[110] Landed society and economy in the central valleys of Oaxaca, however, took a rather different form to the rest of colonial Mexico, in that the sizeable indigenous population managed to retain possession of land in the valley area. All sectors of the population in the region had increased in the late seventeenth century and throughout the eighteenth, but unlike other regions of colonial Mexico, there was much less intermarriage between Indios and non-Indios. This meant that, relative to other regions, the indigenous identity remained strong in both rural and urban settings in Oaxaca (Taylor, 1972: 34). In 1790, for example, 88 per cent of the total population of 410,618 of Oaxaca was indigenous, the Spanish representing only 6 per cent of the total population at this time.[111] Although urban growth in the Valley of Oaxaca proceeded relatively slowly up until the middle of the eighteenth century, Antequera witnessed a substantial increase in the urban population, reaching 18,000 by 1792 (Chance, 1976: 610). Indio populations, however, again made up approximately 28 per cent of this urban population, while mixed race or Mestizo populations accounted for 37 per cent. The equivalent figure for the Spanish population in contrast stood at only 1.5 per cent. Thus, by the close of the eighteenth century, even in the urban areas of the province, the indigenous population remained numerically strong.

The valleys' indigenous pueblos and nobles generally managed to maintain control of sizeable tracts of land. Largely because of the high transportation costs to Mexico City and the wealthier mining districts of the north of the country, there was little in the way of extensive production of staple crops on the Spanish haciendas of the region (Baskes, 2005). As a result of this, together with the strength of the Indio's presence, indigenous communities managed to maintain control over most of the landed resources of the region. By the eighteenth century, the indigenous populations of Oaxaca still possessed the majority of the land in the form of communal landholdings and it was this feature that distinguished Oaxaca from other regions of colonial Mexico. Much land remained in the hands of the hereditary caciques, as a result of their power and status prior to the Conquest, but also due to the role they played in the peaceful transition to Spanish domination, supporting the Spanish administration, acting as tribute collectors and aiding the congregación efforts of the early colonial period.[112]

Baskes (2005) has argued that the retention of land that Oaxacan peasants enjoyed may have also been a function of cochineal's importance in the area. By 1793 there are thought to have been around 25–30,000 people, or 8 per cent of the population, employed in its production (Baskes, 2005: 12). This certainly had important social and economic implications. Few economic possibilities derived from investing in the land could compete with the cochineal market. The land thus remained largely under the control of the indigenous communities (Baskes, 2005).

In the central valleys of Oaxaca, local communities were also 'notoriously litigious' (Taylor, 1972: 82), and literally hundreds of lawsuits indicate that they defended their rights, their lands and their water resources from usurpation legally and also occasionally by force.[113] There are also a great many land disputes between indigenous communities, many of which continued over several decades and, in some cases, generations, as shall be illustrated in chapter 5.[114]

Even towards the close of the eighteenth century the non-indigenous population owned less than half of the total area of the central valleys of Oaxaca, most of which was grazing land. Spanish landholdings per se tended to be fragmented into a large number of small landholdings and ranches rather than large estates. Moreover, few estates were entailed and as such there was a rapid turnover in ownership. Indigenous communities in contrast owned about two thirds of the agricultural land in the region in the eighteenth century and also had a number of important grazing sites. Caciques, particularly in the Etla and Zimatlán valleys, if less so in Tlacolula, leased properties to the Spaniards and indigenous communities (terrasguerros) for a fixed rent. Some of the rental periods were specified, others were indefinite, ensuring a perpetual rent for some caciques.

Mercedes registers indicate, however, that control of land did not always ensure control of the use of water in that land. As figure 3.5 suggests, mercedes specifically for water in this region were not that common, and became less so throughout the colonial period. The majority were awarded in conjunction with milling facilities. Spaniards, caciques, the clergy and indigenous communities were all awarded mercedes for this purpose in the sixteenth and seventeenth centuries. By the eighteenth century, however, by far the majority of awards that were made were to Spanish residents, and none appear to have been awarded to caciques, perhaps suggesting a decline in their power at this time (Fernández-Tejedo et al., 2004).[115] Given water was essential for all kinds of commodity production, this was to have profound repercussions for the environmental security of indigenous communities.

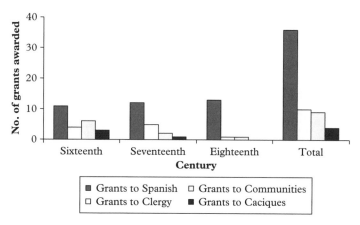

Figure 3.5   Recipients of Mercedes for water use in Oaxaca, sixteenth to eighteenth centuries (AGN, Mexico City Ramo Mercedes)

Climatic Variability and Vulnerability in Colonial Mexico: A Preview

The regionally idiosyncratic environmental circumstances, the economic trajectories, demographic transformations and distinctive land use systems that evolved in each of these case study areas during the colonial period were to play a critical role in determining how populations in each region were affected by and were able to adapt and respond to climate changes and weather events and their implications at different points in time. As demonstrated in chapter 1, the way in which a society (or sectors of that society) can be affected by and is able to respond to and recover from the effects of such changes and events is very much determined by prevailing demographic, social and economic conditions, and community behaviour and practices (Liverman, 1990). There are often differential impacts and losses, depending on the levels of preparedness of different sectors of the population. In turn, preparedness is dependent upon the social and economic status of a particular group of people, and the degree to which they can effectively buffer themselves against the impacts of an event. A crisis only becomes so, however, when a significant proportion of the population is at risk or significant life and economic losses are incurred.

Investigations of the colonial archives for each region have revealed numerous references to droughts, floods, frosts, strong winds and storm events, as well as other natural hazards, throughout the colonial period (Endfield et al., 2004a, 2004b). Although the archives necessarily record those events that were in some way regarded as noteworthy because of their effects, not all of these events necessarily resulted in crisis or upheaval or stimulated significant social adaptation and response. The relative impacts of these events, and hence the degree to which society was able to respond to them, depended very much upon the time and place specific context in which they took place. The precise nature of the impacts and responses associated with some of these events form the focus of the next three chapters, but it is useful to first outline a brief chronology of events as recorded in the colonial records.

Drought-related harvest losses and food scarcities are recorded in Guanajuato, and the Bajío more generally, in 1590/1, 1601, 1641, 1647, 1651, 1662, 1696, 1751, 1754, 1755, 1768, 1784–6, 1793, 1803, 1809 and 1816.[116] Although relatively few droughts appear to have affected society in Oaxaca in the wetter south of the country, late or non-existent rains and water scarcities are documented in 1550, 1690, 1696 and 1726, while harvest problems appear to have descended on the region in 1733, 1746–7 and 1785.[117] As is to be expected, however, drought was a more frequent phenomenon in the arid north of the country. Records from Chihuahua, the majority of which date from the eighteenth and early nineteenth centuries, for example, indicate that drought resulted in harvest failures and food scarcities in the region in 1724–7, 1739–41, 1748–52, 1755, 1758, 1760–5, 1770–3, 1785–6, 1804–6, 1809 and 1812–14.[118] With the exception of successive droughts in the 1770s and 1780s which have been recognized elsewhere in the country, archival evidence of prolonged drought supports tree ring evidence of two particularly severe phases of prolonged drought in northern Mexico in the middle of the eighteenth century (1750–65) (Dias et al., 2002) and between 1801 and 1813 (NOAA International Tree Ring Data Bank, cited in Liverman, 1999: 101).

Relatively little attention has been paid to the historical record of flooding and its impacts in Mexico (Jáuregui, 1997), yet documentary evidence indicates that damaging and, in some cases, catastrophic flooding represented a significant and in some cases persistent threat to society. In Oaxaca, seasonal rains as well as unusually heavy rains regularly led to flooding, especially at the close of the rainy season (September/October), disrupting livelihoods of communities and individuals in these locations.[119] Devastating flood events, causing massive livelihood and life loss, are also recorded on a number of occasions, most notably in 1599, 1721 and 1788. Literally hundreds of floods are also recorded in the colonial archives of Guanajuato, Celaya and Léon.[120] Not all relate to extreme or unusual rainfall. The construction of many earthen and brick built dams, reservoirs and water diversion systems, particularly in the second half of the eighteenth century, may have greatly increased the vulnerability of communities and landowners in the region, especially those located alongside water courses, to climatically induced variations in water availability and water level. A series of particularly severe and devastating floods was recorded in the region in 1692, 1750, 1753, 1760, 1770, 1771, 1772, 1788 and 1804.[121] Some of these events, as shall be shown in chapters 4 and 5, may relate to unusual weather events, but the level of water management in the region, though representing in itself an adaptive response to the unpredictable rainfall, might have in fact exacerbated flood risk, an issue which will be discussed in chapter 5.

Communities across the country were also affected by a number of epidemics during the colonial period. Some of these episodes are particularly well documented (Cooper, 1965). The epidemic referred to as Matlazahuatl, for example, affected a large portion of the central region of the country between 1736 and 1739. A quarter of the population in Michoacán is thought to have been lost over this period,[122] and there are reports of towns across central Mexico being abandoned or 'destroyed' following the epidemic.[123] Similarly, the epidemics of 1761–2, 1797–8 and 1803–4, though well documented for Mexico City (Cooper, 1965), as shall be illustrated, appear to have had widespread impacts on the populations of other regions of the country. Links between harvest crisis, famine and disease are far from new (Malvido, 1973) and it is well known that some infectious diseases can be aggravated by malnutrition as well as population mobility (Kovats et al., 2003). Documentary sources indicate, however, that a number of the reported local and more widespread epidemics may be associated with agrarian problems that may relate to climate events. Figure 3.6, for example, summarizes possible relationships between reported drought related harvest problems and archival references to disease in epidemic proportions across colonial Mexico.

Knowledge of drought, flood and disease events in the past can condition how society comprehends and responds to the problems of uncertainty and how that society not only conceptualizes the likely risk of events, but also anticipates the impacts of those in the future (Koselleck, 1985). In this way, experience or knowledge of climate changes, events and their impacts in the recallable past can effectively become part of the cultural and infrastructural fabric of thought, discourse and practice of a society or community. Weather events or their implications can thus prompt a variety of remedial or mitigating actions, coping strategies and adaptations (Hassan, 2000). The following chapters explore the manifestations and social and cultural interpretations of and responses to some of the recorded weather and weather related events across colonial Mexico noted above, drawing on archival information from the three case study areas. Notwithstanding the level of disruption and, in some cases, devastation caused by the impacts of climate variability, it will be shown that weather or weather related events might have served a number of purposes. Experience and social memory of climatic variability, weather or weather related events and indeed, periods of subsistence crisis and disaster,

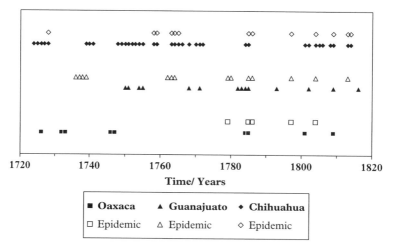

Figure 3.6   Drought, harvest crisis and recorded episodes of epidemic disease in Oaxaca, Guanajuato and Chihuahua, 1720–1820

improved the level of climate knowledge among affected societies and increased public awareness of the differential vulnerability to extreme events and natural hazards. On the one hand, this may have facilitated solidarity among distinctive social groups, as well as a variety of noble, selfless and benevolent acts. Moreover, experience and awareness of climatic variability fuelled a variety of adaptive strategies, innovation and agrarian experimentation. The same events and crises, however, might have equally served to reveal and highlight tensions between social groups. Weather events and crises might have thus contributed to both conflict and co-operation. Exploring such divergent responses provides insight into the ways in which different sectors of society conceptualized, addressed and articulated knowledge about climate change risk, causality and vulnerability and forms the subject of the following two chapters.

By comparing multiple archival references, it is possible to identify that some of these periods of weather or weather related events do appear to have been experienced in all three regions of the country and to have resulted in significant disruption and social and economic crisis (Figure 3.6). The majority of recorded agricultural crises and famines in Mexican history can be shown to be associated with successive droughts, or drought combined with other unusual or extreme weather events, such as frosts (Florescano, 1980). The most devastating impacts resulted where periods of inhospitable weather or weather related events compounded other social or economic difficulties, specifically those associated with population pressure, resource stress and periods of political instability.

Chapter 6 considers how in a context of population pressure, resource stress, inequality, and political and economic instability in the 1690s, on several occasions during the eighteenth century and particularly the later part of the century, a number of phases of repeated or compound weather and weather related events may have triggered a series of nationwide subsistence crises. It will be suggested that although agrarian problems and crises had of course occurred many times before in Mexican history and indeed prehistory, the unique combination of social, economic, political and climatic factors around the late eighteenth century and first two decades of the nineteenth century might have contributed to popular insurgency when previous periods had not.

Chapter Four

Responding to Crisis: Vulnerability and Adaptive Capacity in Colonial Mexico

Introduction

Coping with unexpected situations can involve a complex sequence of mechanisms, behaviours and actions.[1] Coping in the face of adversity, for example, can be seen as consisting of a series of discrete or concurrent, reactive, proactive or anticipatory activities, which can be operationalized at a variety of temporal scales and at a range of levels (Smithers and Smit, 1997). Adaptive capacity, in contrast, refers to the ability of a system, individual or society to undertake these actions in order to adjust to or moderate potential damages, to take advantage of opportunities, or to deal with the consequences, and implies a somewhat longer-term and possibly more developmental and opportunistic process.[2] Adaptation might thus be manifest in the establishment and implementation of trading relationships and market exchanges, and might be achieved through individual or institutional action and decision-making, through legal and legislative intervention, regulation or technological change. It might also involve the marshalling and extending of social networks and relationships (Smit et al., 2000; Adger et al., 2005).

A diversity of factors, however, can influence both coping and adaptive capacity, as well as the actual nature of the adaptive response or coping strategies adopted by a society in a given situation. These are a function of the nature of the perceived threat, and a variety of time and place specific environmental, social, economic, political and demographic conditions and drivers (Adger et al., 2005: 76). It follows that the ability to cope with and adapt to climatic variability and extreme weather events varies over time and according to different contexts.

In recent years there has been a growing interest in understanding society's ability to respond and adapt to climate change, with a particular focus on exploring social capital responses as a form of adaptation (Adger et al., 2005). Indeed, the concept of social capital has become pivotal to many theories regarding adaptive capacity and management in the context of climatic risk (Adger, 2003: 390). Concerned with 'the relations of trust, reciprocity and exchange, the evolution of common rules and the role of networks' (Adger, 2003: 389), it represents a civil society response to coping with environmental risk and as such provides insight into the mechanisms of collective rather than individual adaptation. To date, most social capital oriented studies have tended to focus on collective responses to contemporary or future rather than historical climatic changes. Yet knowledge of successes and failures in adaptation of a society or community to past climatic variability can contribute to our understanding of the capacity of that society or community to respond to environmental threats and climate changes at a range of scales (Tompkins and Adger, 2004). The colonial archives of Mexico provide insight into some of the different innovative coping strategies and social capital responses and adaptations that developed over a range of timescales in the different regions of the country. This chapter will demonstrate how the impacts of seasonal and inter-annual climatic variability, extreme or unusual weather, or weather related events in Mexico prompted the adoption of a wide variety of traditional as well as novel remedial, preventative and mitigating actions, coping strategies and adaptations. All can be considered to be important cultural responses geared towards reducing vulnerability and improving societal resilience.

Moral Economic and Institutional Responses to Climate and Crisis in Colonial Mexico

Many adaptive strategies in pre-Hispanic Mexico were designed to buffer societies against the implications of unreliable rainfall. Irrigation and the development of quite complex and sophisticated systems to manage, store and transport water represent obvious examples in this respect. Periods of agrarian crisis also spurred a range of non-economic, sometimes reciprocal, relationships between patrons and clients, rich and poor. Such moral economic responses offered, in Scott's terminology, a 'subsistence ethic' in that they were intended to provide a minimum level of subsistence and food security during times of want. The organized storage of grains, food products and other saleable assets, for example, represented a vitally important buffer against expected seasonal shortages as well as more prolonged periods of hardship (Blaikie et al., 1994: 66). Farmers traditionally kept (and still keep) seed and grain[3] harvested from their own crop from one season to the next (Badstue et al., 2006), while there are thought to have been a number of formal and more informal indigenous social institutions and networks designed to distribute both food and grains during times of scarcity. The Aztec empire, for example, distributed maize from central granaries during periods of harvest failure and food scarcity. These stores and trading networks, however, proved to be insufficient during the most severe times of dearth. The protracted drought of 1 Rabbit (1450–4), for example, was so severe and prolonged that more desperate survival strategies were adopted. People 'sold' themselves and their children in labour and slavery (Hassig, 1981) and resorted to so called 'famine foods' and liquids derived from cacti, agave and mesquite fruit as a basic form of nutrition.

Many of these adaptive and coping strategies continued to operate in one form or another after conquest. The practice of maintaining a reserve of emergency food supplies, normally grains, to compensate for times of harvest loss, for example, continued into the colonial period (García-Acosta, 1993). As might be expected, Mexico City boasted some of the finest of such 'social services' in the whole of the country. There were government sponsored grain markets or Reales Alhóndigas (royal grain stores) and also granaries (pósitos). Alhóndigas and pósitos were also established in other principal cities across Mexico, but specifically in mining areas and close to the ports of the country, those areas that represented the 'engines' of the colonial economy and where populations were dense. During years of poor or failed harvests, the authorities, as Cope (1994: 75) suggests, 'fearful of mob violence', stocked these city granaries as a matter of priority. There were very few grain stores in rural regions, though it would often be the rural communities that would face the most significant hardship during periods of harvest crisis. Indeed, the lack of reserve food stores in the countryside is thought to have been one factor that contributed to the strong rurality of dissension in the late eighteenth and early nineteenth centuries in the years leading up to the independence movement (Hamnett, 2002), an issue which is revisited in chapter 6.

Actual experience of serious harvest failure may have stimulated the establishment of a number of these facilities in the case study areas. The alhóndiga in San Felipe el Real de Chihuahua (figure 4.1), for example, was founded in the early 1730s after a period of worrying harvest failures in the middle 1720s. The harvests of 1724 had been thin and maize shortages and scarcities of wheat flour were recorded around this time.[4] Council officials had already begun to ration bread,[5] and special arrangements had been made to slaughter extra livestock to feed the poor.[6] This form of 'social storage' appears to have been relatively common practice during the most serious periods of food scarcity and dearth in livestock ranching areas.[7] Despite such measures, however, some of the poorest sectors of society resorted to begging for food from house to house (Martin, 1996: 25) and a public granary was deemed essential in an effort to avoid such hardship from being repeated.

Similarly, in Oaxaca in 1729, preceding a period of grain scarcity that would affect the whole of the region in the early 1730s,[8] requests were made

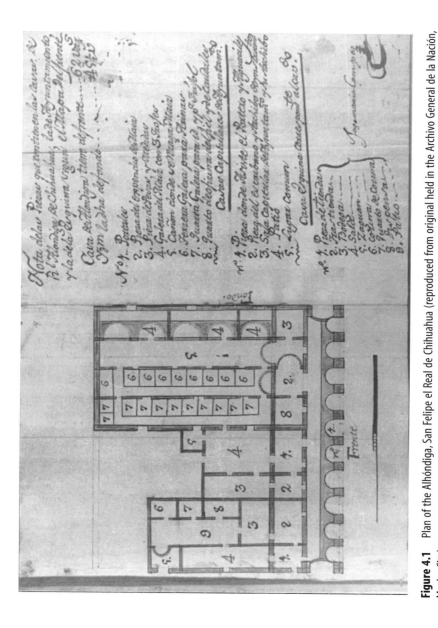

Figure 4.1 Plan of the Alhóndiga, San Felipe el Real de Chihuahua (reproduced from original held in the Archivo General de la Nación, Mexico City)

to build an alhóndiga 'like those they have in all the cities'. The lack of a public granary hitherto had meant that there was 'almost always scarcity experienced in the city (including the present year [1729] when prices have risen to 12 pesos per carga'.[9] The fundamental need for a public grain store would, it was hoped, bring an end to the 'the hardships that the poor people suffer'. Juan Bautista, one of the supporters of the idea, suggested that 'all the other cities had one [an alhóndiga] and Oaxaca did not, and they didn't want it to be said that this was because of poor governance'.[10] There were similar concerns in Celaya a few years later. A note from 1744 indicates that the town at that point had no grain store, yet this was considered to be 'the only remedy to contain the price increases that happen ... with the scarcity of rain or extemporaneous frosts'.[11]

To some extent, storage of foodstuffs through the alhóndigas was a relatively successful strategy to hedge against food scarcity, high prices and hunger. But, as had been the case in the pre-Hispanic period, these stores provided little succour during the most severe and prolonged years of agrarian crises. In Chihuahua in 1758, for example, when drought, hunger and disease were taking their toll on the population of the area, it was recognized that 'because of the droughts experienced in all ... areas, the transporting and trading of seeds and flour to the Real Alhóndiga of the Villa (of Chihuahua City) has also dried up'.[12] Another note from December 1760 highlights the local council's concern over the 'scarcity of seeds which appears to have been presented in the Real Alhóndiga' in Chihuahua.[13]

Florescano (1976, 1980) has illustrated how such variations in maize (and other grain) prices in the sixteenth and seventeenth centuries may have been a function of drought, suggesting that many price rises were preceded by periods of severe drought induced harvest failure and consequent scarcity in food supplies. It seems such links can indeed be identified in the case study areas of concern in this study. In Chihuahua in 1764, for instance, after the effects of drought had severely reduced the harvests, the small quantity of maize that existed in the Real Alhóndiga was described as being of 'low quality' but was nonetheless commanding a 'very high price'.[14] Similarly, the scarcities that prevailed in Chihuahua during the first three (drought) years of the 1770s dramatically affected grain prices.[15] There was only 'a small quantity of maize available in the Real Alhóndiga, its quality low and its price very high'.[16] A document compiled in 1790, but relating, among other issues, to the price increases of the early 1770s, also records the 'great alteration in the prices of these indispensable basic comestibles'. Indeed, the prices rose to an exorbitant 9 pesos per fanega of maize arriving for sale in the Real Alhóndiga in 1770, 1771 and 1772, which caused 'considerable problems for the community and especially the poor and destitute'.[17] In this way, harvest scarcities disproportionately affected the poorest sectors of society, and invariably this meant the indigenous groups.

Speculation and Scarcity: Capitalizing on Climate Knowledge

The colonial system allowed the larger, wealthier landholders, hacendados and merchants to actively manipulate the price of food staples during periods of dearth (Liverman, 1990). They would often be able to capitalize on grain scarcity and desperate levels of hunger by charging exorbitant prices and some may have even withheld supplies to artificially create shortages even though grain might be abundant, so ensuring high returns and greater profit margins. In the Bajío, for example, there was deliberate speculation on the part of some producers and traders who regularly took advantage of knowledge of drought induced harvest failure and grain scarcity to artificially inflate prices. An Ayuntamiento document compiled in Celaya in 1744, for example, highlights how speculation could result in soaring prices even 'without the motives' of drought or frost induced harvest loss.[18] Speculation and profiteering seem to have been particularly problematic around the middle of the eighteenth century when some traders were buying large quantities of grains, occasionally from outside the immediate area, with a view to creating artificial shortages and, in so doing, deliberately pushing up the market price. A document filed on 30 May 1755, for example, suggests that certain individuals had been purchasing maize in Léon 'to take outside of the jurisdiction and others to convert it into their own gains, this causing grave damage to the public' and 'resulting in an alteration of the prices'.[19]

Price regulations and rationing of grain purchases had been implemented in Léon in 1756,[20] and by 1772 similar measures had been introduced in Guanajuato.[21] By this stage, however, the local administrations in both Léon and the city of Guanajuato were growing more concerned about the shortage of grain and general hunger due to 'the malice of many hacenderos who are not content to sell well the products they raise, that they also buy those that others reap and these they deposit in their barns … and they are able to sell them at a price they like'.[22] Nevertheless, it was perceived that this persistent problem

> provided a very good opportunity to construct a store (posito) which … the cities of Castille used in order to avoid shortages of fruits/goods and poverty … and this was one of the best methods that they could follow … for they always maintain the best price for necessary foodstuffs … and they will be very useful in the towns that frequently suffer want and poverty in Guanajuato.[23]

The establishment of a posito in Guanajuato in 1774 indicates that the proposal may have met with general approval.[24]

The problems of speculation and profiteering were not restricted to the Bajío. It was not unusual for traders in Chihuahua to similarly buy produce outside a region at a lower price and to bring it to sell locally at prices that were described as 'very excessive and very damaging to the public'.[25] In 1768, one trader, Don Fernando de la Torre, was accused of illegally holding back his maize in order to secure a higher price.[26] To counter the effects of this opportunism, the local council resolved to also 'bring maize and flour (wheat) from large distances into the province for the Real Alhóndiga so that the public would not suffer from shortages'.[27]

One of the functions of grain stores was to theoretically guard against such profiteering and to assist in price regulation. Where a Real Alhóndiga existed, all grain merchants were officially obliged to sell their produce first through the granary store in order to ensure that a fair price was then offered to the public. All the maize distributed by the alhóndiga on a given day was sold at a set price agreed upon by individual sellers and reflective of prevailing market conditions (Cope, 1994). There was also legislation which ensured that 'no miller should buy or receive flour into his house without first giving a count to the Alcade of the Real Alhóndiga or the judge on duty'.[28] Moreover, it was decreed that 'the farmers and merchants who bring wheat, flour or barley to the alhóndiga ... are not able to retain [the goods] for a time of more than twenty days without selling it.' Indeed, if they withheld grains for longer than this, in order to artificially create scarcity and so raise prices, then they could be ordered to sell it at a specified price.[29] Despite these measures, deliberate forcing of grain prices, competition and insider dealing between merchants and the monopolization and retention of goods remained a key problem into the later part of the eighteenth century.[30] Council minutes from 1782, for instance, suggest that the city of Léon 'was suffering such great scarcity and need of seeds of maize and the rest of the comestible fruits produced by the farmers of the jurisdiction'. Apparently, local merchants and producers 'were retaining the commodities until they reach high prices'. There was thus a need to centralize the produce in a central store in the city 'for the public good'. Local producers were expected to supply maize and flour for the store to sell in normal months and 'not to retain it in the months from June to November' so that prices could be controlled.[31] By this stage, however, the scale of illegal trading had also made it necessary to employ a 'flying guard' in Guanajuato to 'keep a watchful eye over the maize and flour introduced by the ... many extraordinary route ways that the city has ... and to exact a new tax' from those merchants bringing in quantities of maize from outside the area via these routes. Extra 'flying guards' were employed to monitor incoming grain supplies in 1789.[32] Thus it seems that alhóndigas and pósitos appear to have been only partially successful in addressing food scarcities, price speculation and profiteering on the grain market.

Trade in Grains: Providing for the 'Engines' of the Colonial Political Economy

The provision of food to service the mining centres in central and northern Mexico was a high priority during periods of calamity. Minerals represented New Spain's most important export commodity for much of the seventeenth and eighteenth centuries and, as such, were the mainstay of the colonial political economy. For this reason, every effort appears to have been made to ensure food provision and supplies for communities and indeed livestock employed in mining areas during the periods of scarcity. As the breadbasket of the economy, the Bajío was a region to which the government could and indeed did turn for assistance when harvests failed, often, it seems, at the expense of the interests of local populations. This type of prioritizing, of course, was not universally appreciated. Failure of the harvests in Léon in 1601, following the combined effects of frost and drought, led the residents of the town to protest about the transport of much needed grains from the town to service the mines at Zacatecas. In a complaint forwarded to the Alcade Mayor, Juan Alonso, a group of residents argued that one Juan de Montalvo had been purchasing maize in Léon to trade in Zacatecas, at a time when there was 'notable hunger and need which was increasing each day' in their own jurisdiction.[33]

The Bajío remained an important source of grains even when more widespread harvest crises led to scarcity across the region itself. In 1692, after a calamitous year that would result in grain riots across central Mexico, produce was sought for the populations of Mexico City from Celaya, Salvatierra, Silao and the Villa de Léon (Cope, 1994). Special purchases of 45,000 fanegas of maize were made, financed by a donation from the capital's wealthiest silver merchants,[34] though these imports ultimately did little to avert the problems of food scarcity for everyone in Mexico City. On other occasions there were severe problems of food provision in cities actually within the Bajío. Following harvest failures in the vicinity of Guanajuato in 1782, however, individual landowners who could potentially help alleviate public suffering were identified. The Alcade Mayor of Guanajuato, Celaya, San Juan del Rio and the Corregidor of Valladolid sought assistance from the Conde de Rabago who at that time owned haciendas in the Valley de Santiago. It was suggested that he could perhaps provide 120 cargas of wheat to assist with the 'urgency' then faced in Guanajuato.[35]

By the middle of the eighteenth century the organied trade in foodstuffs from region to region and the purchase of grains by local government authorities from less affected areas to supply more hard-pressed communities and populations had become an important coping strategy for drought afflicted populations. Even during what would unfold into the disastrous and widespread famine of 1785–6, for instance, it seems that the wheat farmers of Celaya were able to produce a bumper crop which helped alleviate the suffering of communities elsewhere in the region.[36] A year later, however, food had to be sought from further afield. By the summer of 1786, the council of Léon was scoping the possibility of transporting in maize and wheat from haciendas and towns up to 70 leagues away. Reconnaissance trips had been made to the hacienda of Rincon in the jurisdiction of Zamora, 40 leagues away from Léon, to the town of Penjamo, some 20 leagues away and to Xalapa, which was a distance of some 22 leagues.[37]

Interestingly, this period of crisis does not appear to have affected society in Oaxaca as dramatically. Indigenous land retention and less reliance on wealthy landowners for food provision, together with revenues from the cochineal industry in this region, gave a large proportion of the population more independence and subsistence capability compared to other regions. Indeed, in some parts of Oaxaca the grain harvests in 1786 were particularly good. The Alcade Mayor of Teococuilco, for example, reported in 1786 that they had been 'free of all scarcity' (cited in Florescano, 1981 II: 570, 572), while surplus production in some places was used to alleviate problems of scarcity elsewhere. A letter dated 9 February 1786 from a Señor Narciso Muñoz reports on the 'very abundant harvest of maize' reaped in Teotitlán, Oaxaca. It was felt that 'this great abundance might partially help alleviate the scarcity experienced in the surrounding jurisdictions that had not experienced equal luck' (cited in Florescano, 1981 II: 570).

The purchase of foodstuffs from less affected areas was a particularly important coping strategy for the regularly drought afflicted communities in Chihuahua. Thus, in the 1750s, when there was scarcity and hunger across the entire region, maize was still available for purchase in some isolated locations, including the nearby valleys of Basuchil, San Buenaventure and Carmen.[38] The local authority once again purchased maize from grain surplus areas in 1786 and 1787,[39] though the scale of the subsistence crisis in these years meant that additional quantities of grain, seed and flour were also transported from much larger distances to supply the Real Alhóndiga in Chihuahua.[40] Maize from as far away as El Paso del Norte, 200 miles further north, was imported and sold, albeit at high prices, in the town at this time (Martin, 1996: 26). Following a period of harvest loss in 1804, and grounded in fears that the population could not be sustained until the next harvest was reaped, temporary relief was again sought by purchasing extra grains from the Valley of Basuchil, where problems of the same magnitude had apparently again been avoided.[41] Nonetheless, shortages of maize were recorded in the alhóndiga in Chihuahua City only two years later in January 1806, following another poor harvest the previous year. This re-enforced the need to 'procure from outside'.[42] Similar strategies were adopted between 1812 and 1814 when the coupled impacts of drought and disease descended on Chihuahua and once again resulted in what, by this stage, were referred to as 'the inevitable scarcities … in the alhóndiga'.[43]

Tribute, Food Aid and the Supernatural: Appealing to a Common Sense of Loss

Indio republics had their own community-based strategies for dealing with periods of harvest shortfall, food scarcities and other unforeseen problems. Certain sectors of a community were expected to pay a tax into a 'common fund'. In 1780, a survey of the goods, profits and funds held by each Indio town in Oaxaca was conducted on the instruction of the Governor and General Administrator of Oaxaca. The survey suggests that only married couples were expected to pay into these funds and that the amount they paid appears to have varied from year to year according to the well-being of the community as a whole. Factors such as harvest failure could lead to lower expectations. Interestingly, however, the survey reveals that at this time a number of communities were claiming that either a 'lack of land' or a 'lack of water' had depleted their harvests, thus reducing their capacity to contribute to the common fund. Indeed, a number of communities, like that of San Jacinto de Chilateca, had only managed to raise 18 pesos, with no profit whatsoever, from the sale of their crops for the entire year, the reason being the severely reduced harvests due to 'a lack of water'.[44]

Among the more customary immediate community reactions to periods of harvest failure consequent upon unusual weather was a request for suspension or relief from tribute payments. Crops failed across the Bajío in 1651. In Léon, this 'very sterile' year led to harvest shortfalls. Communities were struggling to reap what they could from their land, but 'there had been no harvest and they could not make any tribute payments or sustain themselves'. Accordingly, they requested tribute relief. The immediate response from the cabildo was the suspension of tax payments for thirty days, though it seems after this time had elapsed the residents were expected to make the payments within six days,[45] a request that in fact may have been impossible to fulfil under the circumstances.

Epidemics, too, of course directly affected the size of the able tribute paying population, and also indirectly influenced the ability to fulfil tribute demands by diminishing the labour available to ensure adequate production of tribute commodities. Between 1762 and 1763, for example, an epidemic of typhus and smallpox swept through the country. According to Señor Baltazar de Viudarre, speaking on behalf of the governor, commoners and naturales of the town of San Francisco del Rincon and the governor of the town of

Nuestra Señora de la Concepción in Chihuahua,[46] this 'common and great epidemic had been experienced in the kingdom', causing very many people to suffer. Because of the 'grave need', the general shortage of food, and the deaths that had occurred in the two towns from the month of January 1762, tribute demands were becoming impossible to meet. The size of the tribute paying population had been reduced and the number of people available for work in agricultural production had also been affected.[47] The remaining tribute population sought relief from their usual payments. Similarly, between 1779 and 1780, many communities had again been affected by epidemics of smallpox and typhus (Cooper, 1965), so reducing the labour pool, which in turn may have affected the ability of communities to produce either tribute or subsistence crops. This may provide another explanation for the difficulties that communities such as of San Jacinto de Chilateca, mentioned above, faced in contributing to the common fund at this time. Similarly, on 27 October 1796, because of the 'epidemic of hunger and pestilence among the livestock' experienced at that time in Léon, the Junta of the hacienda cancelled the remissions of the communities of Coecillo, San Francisco and Concepción de Rincon, despite the fact that they apparently still owed tribute debts from 1789 and the first four months of 1790.[48]

While, for the most part, local government, therefore, appeared to be relatively sympathetic to such requests for tribute relief, it has been suggested that there was 'very little evidence of organized relief' for the most vulnerable communities, specifically the rural poor during crises (Tutino, 1986: 77). The archives do, however, illustrate a number of instances where government aid and assistance was made available for the poorest sectors of society in specific regions during periods of food scarcity. On at least two occasions during the colonial period, crop failure and food scarcity resulted in disproportionate suffering among the poorest members of the population in the Valleys of Oaxaca. This seems to have stimulated a moral obligation for the local and (in one of the cases) central authorities to address scarcities and ensure provision of some level of subsistence for the poor. One instance was in June 1733 when harvest crisis, and consequently 'an urgent need for relief in the form of maize in the city', was noted in the minutes of the city council meeting that month.[49] Grain stores were depleted, prices rose and a request was made of the local council in Antequera to provide food or financial aid for the poor who faced starvation. An unexpected period of drought in 1746 similarly led to the widespread failure of the maize harvests across the whole of Oaxaca. A deputation of administrators from Oaxaca visited the manager of the grain store on 28 June that year, claiming that suppliers were already speculating that there would be an imminent harvest crisis and so were charging 'exorbitant prices'. The administrators thus pleaded for the manager to release some of the grain from the city store to ease the situation.[50] There was little improvement when four months later, on 19 October, it was reported to the local council that 'the maize has totally been lost with the great lack of rains' that summer. There was such 'an urgent need', particularly among the poor and landless who were helpless in the face of such great food shortage, that financial assistance was provided to the tune of 12,000 pesos.[51] As hopes of a late rainy season diminished a month later, a decree from the Viceroy and Captain General of New Spain, Don Fransisco Germes del Ducasitas, dated 11 November, agreed that 5 percent of this 12,000 pesos should be assigned solely 'for the purchase of maize and to assist the urgent need for the public cause'.[52]

Appeals to the supernatural represented another of the more traditional and collective social responses to weather related calamities in pre-Hispanic Mexico (Hamnett, 2002: 4) and elsewhere (Martinvide and Vallve, 2005), and one that prevailed in one form or another into the colonial period. As had been the case with the variety of pre-Columbian earth and water deities, there were numerous images associated with the Catholic Church which were appropriated by the indigenous populations of New Spain as having power over land, fertility, disease and natural disasters. 'Rogation' ceremonies to these images were carried out with the expectation of bringing an end to environmental or morbid vicissitude associated with drought, harvest problems, widespread epidemic disease and other catastrophic events such as flooding and earthquakes. In Mexico City these ceremonies were dedicated, most of the time, to the Virgin of Los Remedios. A statue of the Virgin would be moved to the Metropolitan cathedral in the centre of Mexico City during periods of crisis, drought and hardship and would remain there until the crisis had waned, rain had arrived or conditions improved (O'Hara and Metcalfe, 1995). For instance, appeals were made to the Virgin in 1618, 1621, 1623 and 1624 when droughts prevailed across central Mexico (Gibson, 1964). Such a course of action, taken during the most desperate of crises, was intended for the common good and often involved deliberations between secular and ecclesiastical authorities in specific urban centres. However, some periods of calamity stimulated more widespread rogation ceremonies across entire regions. A letter dated 22 June 1768, for instance, details the suspension of the rains across central Mexico at this time, which was thought to have 'fermented an illness of measles, though there was not much mortality'. Interestingly, the author of the letter also suggests, somewhat sensationally, that 'the lack of rains is the cause of the illness that is suffered and the repeated earthquakes' experienced from 4 April of that year.[53] While the drawing of such links might reflect a contemporary sensationalism that was not uncommon at the time,[54] it is clear that this devastating sequence of calamities affected societies across a broad sweep of central Mexico. Region-wide rogation ceremonies to the Virgin de los Remedios were held in the hope that her divine intervention might bring an end to the suffering.

The lack of summer rainfall, coupled with further reported earthquake activity in central Mexico in 1771, similarly led to instructions being issued to the whole of the Archbishopric to undertake organized rogation ceremonies once again.[55]

Rogativos also focused on other sacred images. The Virgin de Guadelupe, Mexico's patron saint celebrated in song, verse and paintings, represented a supernatural image of motherhood and nourishment, health and salvation.[56] Her services were also frequently called upon during times of drought, harvest failure and epidemic disease (Lipsett-Rivera, 1999a). Guadelupe was, for example, hailed for bringing an end to the Matlazahuatl epidemic which affected much of central Mexico and which, by 1737, had claimed 40,000 lives in Mexico City and 50,000 in Puebla over an eight month period (Peterson, 1992).[57] Similarly, in Chihuahua, there are many occasions when the populace resorted to her intervention or gave thanks for her apparent intercession. In 1738, for example, offerings were made to the Señora de Guadelupe for preventing a nationwide spread of the food scarcity and famine which devastated parts of Durango and Chihuahua,[58] while in 1755 communities were praying for divine intervention to relieve them from the 'calamity of drought experienced three [years] in a row, up until 1751', though the document indicates that divine intervention had first been sought in 1748 when the drought began.[59]

Divine intervention was sought once again, though this time to bring an end to the 'grave epidemic' that the population of Chihuahua was suffering throughout the 1760s. This had caused 'the death of diverse people, and especially the poor and destitute who lack food'. A rogation ceremony was proposed for 21 March 1764, followed by another to Nuestra Señora de la Virgin Maria, and there were also appeals to the general public to donate goods and money to Brother Don Juan de Rivera, to 'continue the good work in which he was attending the poor and infirm'.[60] Similarly, the widespread harvest crisis in 1785 and that of 1813, which affected Chihuahua in particular, again stimulated repeated appeals to Señora de Guadelupe.[61] On 27 July 1813 the council noted that 'in view of the suspension of the rains ... and the frosts' a special plea for help should be made to Nuestra Señora de Guadeloupe.[62] The same document also details the costs that such a ceremony would incur and how these would be met. A special community tax would be levied to cover the bulk of the expense, while voluntary donations to the tune of a total of 140 pesos were needed, not including the wax for the candles and the rosary, which alone were estimated to amount to 66.5 pesos.[63]

The images of Guadelupe, and the Virgin of Remedios, among others, represented the embodiment of hope, salvation and benevolence for all cross-sections of the community. In Wolf's (1953) words, they provided something of a 'collective representation' of Mexican society during the colonial period and the rogation ceremonies themselves represented a collective response to periods of calamity and crisis, especially where other coping strategies had failed.

It is clear that the shared concerns and values, the common sense of loss during or following agrarian crisis and catastrophic events, and perhaps also contemporary concepts of civic responsibility, might have served as a catalyst for such community engagement, cohesion and co-operation (Riley, 2002). In fact, in some situations, shared environmental concerns might have actually transcended differential class divides and power relationships. Collective action was at the very heart of many decisions regarding the management of natural resources and, as shall be illustrated in the following section, was pivotal to the coping strategies of societies accustomed to living with risk (Zeigler et al., 1996).

'Compadrazgo', Community Engagement and Public Works

Understood to be the voluntary action taken by a group to achieve common interests (Meinzen-Dick and Di Gregorio, 2004), or the 'co-ordinated behaviour of groups towards achieving a common ... purpose' (Vermillion, 2001: 184), community engagement commonly involves several actors, and is usually directed towards achieving a common good (Badstue et al., 2006).[64] In many parts of Mexico, collective action, social relations and networks between communities played, and continue to play today, a key role in averting or dealing with crises. Research in modern-day Oaxaca, for example, has revealed a variety of social relationships and networks that come into being during periods of food scarcity. In addition to biological and affiliate kinships within communities, there are 'fictitious kinships' which might include friendships and 'compadrazgo', that is to say, close social relations based on reciprocity, trust and mutual help (Badstue et al., 2006). Such formal and informal relationships also existed in the colonial past.

In addition to the formalized and institutionally organized trading in grains and foodstuffs during crises, for example, there may well have been more informal social arrangements and compadrazo agreements between farmers to ensure seed acquisition during times of dearth. Moreover, there is evidence to suggest there may have been other forms of collective community action. Riley (2002) has illustrated for Tlaxcala, for example, the marshalling of social networks in the construction of public works. Elsewhere and in other situations, community effort may have been similarly galvanized for the common good.

Indigenous communities in particular played a key role in efforts geared towards ensuring an adequate urban water supply in the settlements, towns and cities of New Spain. By 1647, for example, population expansion in Celaya was placing pressure on a limited water supply. By this stage, four convents had been founded in the area and there had been significant urban population expansion. The solution to urban water scarcities lay in bringing the water to the city via a pipeline from a source located 1 league away. The cost of this project was estimated to be around 14,000 pesos, but it was suggested that the members of the religious orders of Carmen and the citizens of the town co-operate in the project. The wealthy residents of the town were encouraged to sell some plots of land to raise extra funds and it was argued that the profits from the local slaughter-houses could also be invested in the project.[65]

Similar potable water schemes had been implemented in Oaxaca almost immediately following conquest. Indios from Jalatlaco, for example, were responsible for maintaining Antequera's water supply and regularly provided their labour in agricultural investments, as well as policing the weekly market, and also helped with numerous public works both financially via tribute payments, but also in terms of labour.[66] Water harvesting proposals, which were altogether relatively cheap to implement, were also forwarded in 1565, though, perhaps significantly, in order to provide water for the Indio barrios rather than colonial towns where there were more Spanish and Mestizo inhabitants. The Audiencia de Mexico ordered the Alcade Mayor of Antequera, Rodrigo Maldonado, to inform the residents and council of the city to 'gather water from the hill slopes, around the town of Gueyepa to transfer it to the barrio town of Jalatlaco'.[67] A rapidly growing urban population in Antequera in the mid-eighteenth century, however, would nevertheless result in more significant public works projects. Gay reports on the Ayuntamiento's decision to construct an aqueduct which was regarded as 'indispensable' not least because the residents of Oaxaca were drinking 'water of very limited quality' at that time. There was particular need to supply water 'during the months of March, April and May', that is to say, towards the end of the dry season when water was scarce yet was critical for some communities.[68]

Such seasonal variations in water availability may have driven other public works programmes and water schemes, while the need for an adequate, potable water supply may have been made more acute by recurrent periods of drought. This seems to have been the case in Chihuahua. In 1748, at the start of a period of drought, harvest crisis and epidemic disease that would affect much of the north of the country until well into the 1750s, for instance, funding was being actively sought to survey for and investigate water sources in the vicinity of San Felipe el Real de Chihuahua.[69] Not only was potable water needed for the growing urban population, but also for the 'much needed' hospital in the town. This was essential in order 'to avoid the general illnesses associated with the original river because, being small, it carried the transport of metal wastes, it was used to wash the clothes of the infirm, [it was used] as a mud bath for the dirty animals, horses, mules and cattle so that it never arrives clear because it is so full of noxious filth'.[70] Built by Tarahumara Indians from the sierra in the 1750s, in return for payments of maize, meat and salt,[71] the aqueduct that brought water to the centre of the city went some way to resolving these problems (Martin, 1996: 42). Despite these efforts, however, there were still concerns over the quantity and quality of water supplies in the city towards the close of the century.[72] In 1791 Don Antonio Comaduran, a surgeon in Chihauhua, noted 'the very grave damage that was being done to the public health in the months of the dry season from the river which barely runs in some places because of the filth resulting from the washing of clothes of the ill people, the lead waste from the mining haciendas and other things harmful to the health'.[73] Coupled with general concerns over the periodic scarcity of water in the growing settlement, such arguments were most certainly influential in the development of large scale water management schemes designed to introduce clean potable water to the city in the 1790s.[74]

In some cases, drought and water scarcity may have necessitated co-operative efforts between different sectors of the community. This seems to have been the case in areas where demands for water, particularly for irrigation purposes, were high. There had already been substantial water management for irrigated wheat in the Valle de Santiago in southern Guanajuato by the beginning of the seventeenth century. It is known that the Brazo de Moreno irrigation canal and drainage ditches had been constructed along the flood channels of the River Lerma, around the turn of the 1600s (Gonzalez, 1904: 232–233; Murphy, 1986: 65–66). Water was managed effectively between stakeholders in the region, with each landowner in the vicinity of the canal holding rights of access to the water for a ten-day period.[75] As was true of much of the Bajío, the first half of the seventeenth century witnessed an expansion in the agricultural exploitation of the valley area, again specifically for wheat production, though there was also some (possibly non-irrigated) maize cultivation and cattle ranching in the area. By the second half of the century water was divided between landowners according to the 'tanda' system, which consisted of shifts of 'turns' in the use of irrigation water, usually measured in days of water.[76] As was common with such water sharing arrangements, water judges were appointed to adjudicate fair water distribution according to the agreed proportions. Water continued to be distributed relatively efficiently between different users, though there were several attempts to confirm and clarify water rights in the eighteenth century. As Murphy (1986: 78) notes, a water judge was appointed and all landowners in the valley met to collectively decide and confirm in writing rights and days of water access. While there was a clear basis for co-operation over water management in this area, however, this did not prevent some bitter legal wrangling over water rights, particularly in the eighteenth century, a trend that is recognized elsewhere in the region and across Mexico more generally, and which is discussed in more detail in the following chapter.

The drought of 1780, however, appears to have stimulated the development of a collaborative water management project between landowners and the Augustinians in the region. By this stage, the Augustinians were among the largest landowners in the valley and monopolized much of the most fertile stretches of land alongside the Lerma, between the valley and Lake Yuriria. Their estate included the Hacienda de San Rafael and the Hacienda de Bolsa, as well as several smaller properties.[77] They invested heavily in wheat production, particularly on the Hacienda de Bolsa, relying on a system of water provision that is thought to have included the construction of a canal from the south of the region. Water shortages, particularly those experienced during the drought of 1780, posed a problem for society in this irrigation dependent agricultural region. The solution lay in exploitation of the waters of Lake Yuriria. The lake was connected to the Lerma by a small channel and so its water level fluctuated with the level of the water in the river. A project was devised to use the lake as a reservoir by placing, in Murphy's terms, 'head gates' at the entrance to the canal, so enabling the regulation of the flow for when water was needed. The landowners in the Valle de Santiago were charged with raising the sum of 5,000 pesos to finance the construction, while the Augustinians secured rights to the use of the lake from the community of the town of Yuriria. By this stage, water was distributed not on a diurnal basis but according to the amount of land that each stakeholder possessed.[78]

Although the scheme was very much a collaborative venture, it was not without its problems. The Hacienda de San Nicolas, alongside the lake, was flooded for half of the year when the waters rose to a level whereby they became usable for irrigation by the landowners in the Valley de Santiago. In an effort to curtail the flooding while maintaining the provision of irrigation water, the Augustinians allowed the lake to be used as a reservoir only during specified times of year and a dyke was eventually constructed from the lake to drain off the water. In reality, the management of water was not as egalitarian as it might have been and the Augustinians maintained tight control over the level of water in the lake throughout much of the colonial period (Murphy, 1986).

'Most sensitive and saddening events': Flood Risk and Social Capital Response in Colonial Guanajuato[79]

The destruction wrought by sudden, violent and uncontrollable flooding was recognized as a recurrent problem for landowners and businesses alike and represented a particular problem for riparian communities across Mexico. Literally hundreds of floods are recorded in the colonial archives of Guanajuato,[80] and there was an acute awareness of the risks posed by the rivers in this region. The River Laja, for example, was considered to be 'one of the most heavy flowing rivers of the whole of the Americas', a fact borne out by the fact that 'its terrifying waters caused infinite numbers of deaths occasioned by its very many floods'.[81] 'The tempestuous flow of the river [Guanajuato] which runs through the middle of the city' of Guanajuato also posed a similarly frequent problem.[82] One contemporary observer, writing in 1788, suggests that flooding in Guanajuato was a function of the city's surrounding geography:

> The waters of the rains were channelled by the steep slopes and arid terrain ... and they [the people] fear the falling waters and often they occasion significant damage and loss, and that the most sensitive and saddening events have been seen in the last few years when the waters have increased and pour down from the hills to the east and south and bathe the city.[83]

Particular buildings affected by flooding were seen to be at risk because of their location. In a document dated July 1637, for instance, Brother Pedro Guitierrez Gaton, Prior of the hospital-convent of Espiritu Santo in the Villa de Léon, notes that 'there was a flood in the river which passes alongside our convent and which took away a great part of it [the convent]'. A series of impartial witnesses provide evidence to suggest that this was not necessarily a rare or unique event. 'The convent was founded twenty years before, over which time, the hospital had seen the river flood on three or four occasions ... the waters rose, leaving the main channel and inundating the rooms/chambers and patios of the convent and also the church'. Another witness indicates that the problem was directly associated with the location of the hospital relative to the river. 'The hospital de Jesus had suffered much damage because it was located lower than the rest of the buildings ... it had survived other flood events, one in the middle of the night, that was so big that it posed a threat to many people.' This witness, a wealthy local resident, apparently had first hand experience of the flood event. He mentions that he had been involved, along with his servants, in the rescue of some of the more infirm in the hospital, taking them to slightly higher ground, adding that the hospital, in his opinion, was located in a high risk area.[84] Whatever the cause, flooding was recognized as a recurrent problem.

The majority of recorded flood events, however, were ascribed to human manipulation of the water supply in the region and may well have been a function of the complex myriad of water diversion channels, dams and reservoirs that had been developed by the eighteenth century. Dam breaches were commonplace events and, while most instances relate to poor dam maintenance and neglect, in some cases floods may have been the result of deliberate sabotage related to longstanding feuds between neighbouring landowners, a number of which are considered in the following chapter.

Flooding had been recognized as a persistent problem in some locations by the middle of the seventeenth century, but there was a series of particularly severe and devastating floods recorded in the region in 1649, 1692, 1749, 1750, 1753, 1760, 1770, 1771, 1772, 1788 and 1804.[85] The scale of the losses incurred suggested by some of the more dramatic events stimulated a range of immediate reactive responses and coping strategies, as well as longer-term anticipatory flood alleviation and remediation schemes. Indeed, a number of these events remain prominent and became inscribed into the memory of local communities through the apparatus of adaptation, response and recovery.

Following the flood event in Léon in 1649 and, in fact, recognition that flooding was a recurrent problem in the area, a flood commission was established to investigate the level of damage and to try to prepare a post-disaster strategy. Later high impact events stimulated much more ambitious and organized structural flood alleviation programmes in the area. Following the dramatic flood on the Laja at Celaya in 1692 (which opened this study), for example, a cross-section of representatives of the city undertook a survey of the local river systems 'with the objective of … developing measures to conduct water to avoid new floods', finally deciding that

> the only way to avoid flooding was the construction of a dam above the river … to contain the waters that come down from the hills and to release this water, little by little into small storage areas, forming in this way a controlled water flow, so reducing the danger of large floods which have proved so dangerous for the city.

Significantly, the undertaking was heavily dependent upon the good nature, or perhaps shared grievance, of the residents and landowners of the area, though the actual construction of the scheme was very much influenced by class distinctions. While Spanish residents were requested to make financial contributions towards the cost of the scheme, the indigenous residents were asked to provide their support in the form of manual labour. This was intended to be very much a collective effort. It was also recognized, however, that there had been an increase in the level of the river water associated with the build up of silts at the confluence of the River Laja and River San Miguel and a suggestion was made that the residents 'would spend the whole of their lives cleaning out, year after year, the silts from this area to avoid blockages' (Zamarroni Arroyo, 1960). Although followed through to fruition, judging by the number of later flood events it is thought that the project was only partially successful, largely because the local council failed to maintain the dam works (Murphy, 1986: 36).

The flood which affected Guanajuato on 5 July 1760 was another high impact event, and similarly 'resulted in turmoil … causing the disaster to survive in lamentable tradition as one of the most memorable that has been experienced'. The event took place 'at twelve at night and finding the river without a bridge or any flood defences ... the houses were flooded with water, the roads were covered and there were innumerable people drowned or unspeakably injured', and in this sense affected a totally unprepared and hence very vulnerable population. Travelling priest Fransisco de Ajofrín was in Guanajuato at the time and provided a graphic eyewitness account of the events as they unfolded.

> The city experienced great destruction of the houses, haciendas and among the people because of a furious thunder storm which caused a sudden burst of water in the nearby areas (a phenomenon they call here 'snakes of water'), which caused an astonishing increase in the river and the ruination of houses, drowned people, destroyed water falls and caused an infinite amount of damage. (Cited in Moreno, 1986)

The event stimulated a number of structural responses:

> Various bridges were built, they erected many fences, the foundations of the houses and haciendas were re-enforced, and also these and the streets were raised with the same materials as were drawn from the river ... it was resolved that it was important to build up a slag heap, as had been done in times past, to raise the level of the road and houses to a higher level to ensure that the river bed/stream bed is always lower than the level of the houses.[86]

These adaptations do appear to have been partially successful as 'twelve years later in 1772 there took place another flood ... although it caused fewer damages because of the actions and precautions adopted in the previous flood'. This event was succeeded on 27 July 1780 by a flood which was apparently 'greater than that of 1772, but less than that of 1760 in its impacts because the river channel was not without some defence ... this inundation, although big, did not cause any precise damage'.[87] Nevertheless this was still regarded as 'fatal flooding' and the damage was significant. Flood waters 'broke the bridges and the fences, the river taking a precipitous course to the centre of the city'.

Following the 1780 flood, the local council took measures to 'reconstruct from new, the bridges that were destroyed and at the same time the fences/walls, both of which are necessary things'.[88] A document compiled 24 years later in 1804 indicates that the scheme had not eliminated the problem. At this stage, there was still an 'urgent need to build a bridge ... and to construct two big pillars to give direction to the waters and to avoid the floods which this city has experienced'.[89] A separate manuscript charts the need to rebuild another bridge known as Granaditas de Salgado in Guanajuato which had also been damaged in the repeated floods, for fear that the 'river will be intransitable in the next rains because the waters coming off the mountains will be copious'. It was anticipated that the onset of the rainy season after the long dry spell might lead to land slips and massive erosion of material from the slopes into the river, thus increasing the risk of flooding. The bridge 'was not only considered useful, but necessary' to avoid the anticipated flash flooding that might ensue.[90]

Other bridge building projects were proposed following flooding in Celaya in 1781.[91] The inundation particularly affected 'the poor people of the towns of San Antonio … and San Juan which had endured the said floods for three days'.[92] The council first made available 2,806 pesos 'to provide relief' for the affected communities. Then, in 1784, in response to the loss of earnings caused by this apparently recurrent problem, it approved a licence 'to build a bridge on the River Turbio in the jurisdiction of the Villa de Léon for the benefit of the residents'.[93] Elsewhere, there were similar developments. There was an 'urgent need to raise a bridge' in Guanajuato around Camacho in 1796 'to avoid the floods which this city has experienced', though again there were concerns over the financing of the project.[94] A series of documents compiled between 1799 and 1803 suggests that similar measures still needed to be implemented on the Laja River. Because it was necessary to cross the river to trade with the surrounding region it was resolved 'that the bridge should be made more solid, the cost of the construction being 13,000 pesos'. A complementary flood alleviation plan was also proposed involving the construction of a series of 'steps/terraces … to contain the sediments and other materials that could not pass through the eye of the bridge'.[95]

It was indeed widely recognized that the accumulation of sediment and debris in river channels and behind dams designed to store water significantly raised the risk of flooding following even normal seasonal fluctuations in rainfall. Dredging and clearing, therefore, were among the most common flood prevention techniques and several documents refer to the need to clear the rivers of sediment and debris in order to reduce the flood risk. Individual landowners were responsible for ensuring the steady flow of waters that passed through their lands and hence were often charged with the task of clearing their own stretches of the river.[96] In fact this appears to have been a condition or proviso of many water agreements and grants. Yet there were also community and city-based clearance schemes that involved much more participation by all sectors of society. Flooding in Guanajuato in 1749, for example, led to the development of one such dredging scheme, devised and forwarded by local residents and land users, particularly those people living in one street, the Calle de Alonso. There residents had described themselves as 'being in great danger' because of the river. They suggested the main cause of the problem lay in the 'rubbish and waste of the city and the material from the hills and mine works' which was being dumped or washed into the river. It seems there had been previous schemes to clear the river of such debris, but that these had not been maintained.[97] The residents proposed the adoption of additional measures to prevent flooding, which included the construction of flood defences in all the areas that had previously been flooded and an ambitious plan to reduce the steep nature of the ravines/gullies surrounding the area, so reducing the likelihood of material being washed down into the river. Other proposals that they forwarded included diverting some of the water from the channel via a series of dykes or reducing its 'power' through a number of small waterfalls. It is clear from these proposals that here was a group of people bonded by a shared interest and common goal to reduce exposure to further flood risk.

In response, or perhaps as a reaction to this proactive community, the local administration enlisted a group of apparent 'experts' for their opinion on the problem, though it is not clear from which backgrounds or with what expertise they came. They produced a flood risk map that showed the areas where debris had built up and which were most in need of clearing, declaring that 'it was obvious that the river had been cleared in the past but ... it was necessary that this should be done every year and to save the 5,000 peso cost for this procedure, it would be simpler to ask residents to not throw their rubbish and waste into the river'.[98] Expert opinion thus supported the need for regular dredging, but raised the thorny issue of the costs incurred by such a continuous and high maintenance endeavour. Moreover, they effectively attributed the cause of flooding and, by extension, any efforts geared towards reducing the flood risk, to the local community.

Interestingly, additional measures were also suggested for the most vulnerable households, including 'those residents that lived in Calle de Alonso'. Given the disquiet that these residents had voiced following the devastating flood of 1749, it was resolved that flood warnings should be 'given to the people who live there using a brief noise or signal indicating that they must leave the houses and close the doors of the houses'. A list of the buildings perceived to be most at risk was also issued. Another measure, forwarded by the residents, however, involved actually using some of the material dredged from the river to build a flood barrier. There was some resistance to the proposed flood alleviation and prevention schemes. Though the use of a 'slag heap' in this way was thought to be a good idea so long as it was properly and regularly maintained, it was claimed by some local landowners that this 'would require the industry of a multitude of men to first clean the river from the point where it drains out'. There was also concern among some of the residents themselves that the use of this flood barrier would allow pools of water to collect behind the barrier and stagnate, and 'this would infect the air, corrupting it and causing the residents to inhale this'. Aside from environmental health worries, the 'insupportable costs' of the proposal posed an additional problem.[99]

Such expressions of community concern did extend beyond rhetoric. Following a flood in Guanajuato in 1788, there was a well grounded fear that 'the arches of the bridges of the river' were once again posing a risk that could have an effect 'if precautions are not taken'. Yet on this occasion the local council resolved 'to clear the river, the eyes of the bridges and various channels and conduits … so that even though the rains might be abundant, the waters will not leave the main channel'.[100] Anticipating the complaints from local landowners concerned that the financial costs of the task might fall on them, the council suggested that the labour for this task could be provided free of charge, 'the offenders of the local prison bearing the cost of the work'.[101]

Similar schemes appear to have been proposed elsewhere in the region, though funding remained a key problem. In 1792, for example, the council asked the Real Hacienda for 500 pesos 'to pay for the clearing of a bank of sand that is in the river'.[102] Other schemes combined this tactic with more significant changes in urban infrastructure. Indeed, repeated episodes of flooding in Guanajuato had, by January 1805, resulted in a resolution to rebuild the main square, the Plaza de San Diego, effectively 'raising it to avoid flooding'. Nevertheless, it was suggested that the practice of 'clearing of the river by hand' should continue.[103]

Flood events thus stimulated both reactionary and anticipatory responses at a range of levels, from the individual through to the local administrative level. All sectors of society were affected by flooding and all were in some way involved in developing, financing or implementing flood alleviation schemes. The build up of debris in the river channels, behind dams and around bridges was recognized to be a particular problem. Dredging and clearing were thus among the most common flood management strategies adopted and practised. A number of flood alleviation projects involving substantial structural investment also seem to have been proposed and, in some cases, implemented, though the lack of maintenance of these works may in fact have exacerbated problems of flooding in the region. Perhaps most significantly, however, although there were some concerns over the financing of flood alleviation schemes, there appears to have been a degree of community participation and action in flood alleviation. Indeed, flood management and recovery may have engendered a sense of civic responsibility, co-operation and collaboration between different sectors of society, particularly during the second half of the eighteenth century when flooding appears to have become more frequent. It is clear that Indio communities and Spanish residents could and did collaborate and work together to address such common problems.

'Great floods' and 'Strong winds': Damaging Events, Adaptation and 'Non-Adaptation' in Colonial Oaxaca[104]

The Atoyac, Salado, Xalatlaco and Tehuanatepec all provided key sources of water for irrigation and domestic purposes in colonial Oaxaca. Indeed, the fertile alluvial soils of river floodplains were prized for floodwater farming. Controlled use of water from flash flood events, for irrigating crops, was also commonly practised (Kirkby, 1973: 36). There was also extensive use of swamplands adjacent to rivers for seasonal grazing purposes. Unusually heavy rains, and flooding, especially at the close of the rainy season around September/October, however, regularly disrupted the economic livelihoods of communities and individuals in floodplain locations.[105] Flooding of the Atoyac River around the capital city of Antequera posed a particular problem, the scale of which was reported in 1561 when the city was being regularly 'bathed' by floods and the area south of the market square was being engulfed by reeds. In response, it was proposed to 'move the river three hundred meters so that it passed by the foot of Monte Alban', a scheme that was begun in April of 1561. Five hundred Indios drawn from the towns of Cuilapa, Etla, Tlalixtac, Tlacochahuaya, Coyotepec, Zaachila, Huitzo, Ocotlán and Tetipac provided the labour for the project which would, it was argued, benefit the whole city (Gay, 1950: 208).

Yet flooding continued to present a problem for both urban and rural communities in the central valleys, affecting key economic sectors across the region. Salt deposits in Oaxaca, for example, began to be exploited commercially soon after the Conquest, but production was regularly disrupted by periodic flooding.[106] Extraordinarily heavy rains in 1723, for example, resulted in the community of Guadelupe, subject town of Etla, 'losing all the harvest of salt' to rising water levels. In response, the entrepreneurial community petitioned for a licence to build a dam to allow them to regulate water levels and so prevent future flood events and economic losses. Their proposal, however, would also provide them with an opportunity to diversify their economic base. They argued that the reservoir behind the dam would allow the community to develop fishing resources as an economic fallback should a similar situation arise again. As it transpired the project never came to fruition because the downstream farming community of San Bartolomé of the jurisdiction of Magdalena opposed the scheme, amid fears that water storage upstream would detrimentally affect their access to waters which were vital for irrigating their crops.[107]

Salt reserves elsewhere in Oaxaca would be similarly affected by later flood events. Another 'strong wet season' recorded in September 1787, for instance, caused problems for the residents of the town of Juchitan in the jurisdiction of Tehuantepec to the southeast of the central valleys. The unusual weather led to 'a consequent raising of the water levels to such a height that it flooded almost the whole salt store'. It is clear from the document that flooding had previously affected salt production in the town, and for this reason the salt had begun to be kept in stores located on small 'islands'. In this case, however, even this adaptive strategy had proved ineffective. An inspection by a local government official suggested that in this year the water levels were especially high because 'it had rained excessively for twenty days', so flooding even the island reserves. In total, an estimated 733 cargas, two arrobas and five books of salt were lost.[108]

Local authority concern over the degree of economic and, in some instances, life loss associated with such events led to the development of various remedial and mitigating measures in case of future events. A document compiled between the years 1766 and 1774, for instance, charts some of the actions taken to overcome trading disruptions when the Atoyac flooded. 'Because of the damage that the Rio Atoyac and its tributaries caused, the construction of a bridge is proposed … it will be very useful because of the great floods and the difficulties faced by all the traders'. As was the case with many of the bridge building projects in Guanajuato in the eighteenth century, the project was regarded as essential if economic losses and trading disruptions were to be avoided each time the river flooded. Moreover, the construction of the bridge would reduce other problems in the city at this time – crime, vagrancy and overcrowding in the city's jails. A few years previously, the Conte de Revillagigedo had noted the increasing problem of vagrancy in the city of Antequera, calling for a 'remedy' to the number of incoming migrants. One solution he had forwarded was for the vagrants to be settled in two parishes, those of San Juan or Santiago; another was for them 'to be educated and trained in particular professions and businesses as apprentices … for a limited time in a field to which they are inclined'.[109] As with other schemes elsewhere, however, the 'employment' of vagrants and criminals in public works schemes in the area provided one of the most practicable solutions. Thus, it was suggested that 'a licence be sought to apprehend all the vagrants and thieves and put them to work in the construction of the bridge'.[110]

Flooding also encouraged an element of opportunism. Some indigenous communities seized on the impacts of flood events to petition for relocation. A number of settlements were practically destroyed by flooding in the vicinity of San Miguel Alupana, Teococuilco in 1653. In response, the indigenous residents of the area petitioned to be able to relocate 'in a more comfortable location in a part and location of their choosing'.[111] Some individuals capitalized on floods, while, ostensibly at least, benefiting the common cause. One entrepreneur in Oaxaca, for instance, saw the sudden availability of water following the 'continuous rains' and 'excessive increases' in rainfall in the summer of 1802[112] as an opportunity to address a longstanding problem of ensuring sufficient potable water for the urban population. The unusual rains had caused an increase in the level of the waters in the Atoyac River, which resulted in burst dams and breaches in the riverbanks, destroying crops and inundating riparian grazing sites. Previously, as one document suggests, 'the public had to travel close to half a league away in search of water to drink and for other necessary uses'. In order to begin to repair the damage done by the flood and to make use of the abundance of potential drinking water, however, Don Fransisco Vilchis suggested that a series of drinking troughs be constructed to drain away and redirect the waters to the needy homes.[113] Unfortunately, the records disclose little more about this ambitious and ostensibly altruistic project, so it is not clear if it ever came to fruition.

Coping strategies and adaptation depend on the assumption that an event, and its impacts, will in general follow a recognized or familiar pattern and that previous experience of such events will provide an approximate guide on how to cope or respond (Blaikie et al., 1994). In sum, when people are aware that an event might occur, they can prepare for it and draw upon ways of coping with it based on past experience (Douglas, 1985). If an event is unprecedented, however, its impacts can be expected to be correspondingly greater. A series of consecutive flood events thought to be on the Rio Mixteco in the north of Oaxaca, for example, recorded on 11, 12 and 13 January 1804, wreaked irreparable damage on the town of San Jorge Nuchita in Huahuapan. Because the community had apparently 'never before seen an inundation of this order', the impacts were consequently dramatic. Although the people themselves were miraculously unharmed, haciendas and associated living accommodation alongside the river were completely destroyed, crops were devastated and livestock killed. Local irrigated farmlands and orchards were flooded 'leaving only a site like a stony beach, with the exception of the church and some houses which were saved because of their elevated location'. In view of the destruction visited upon them by this catastrophic event, the community submitted a plea for extended relief from normal tribute/tax demands.[114]

Some events have lengthy return periods such that precedents are imperfectly understood and registered. There can, in effect, be a degree of disaster amnesia in that the social memory of the event might not be sufficient to translate into a recognized adaptive strategy. Indeed, in such situations coping or adaptation may not necessarily apply and 'inaction' may in fact represent the main response (Paavola and Adger, 2006: 596). A number of apparently catastrophic, if rare, extreme weather events appear to have stimulated little in the way of adaptation. Two flood events in Oaxaca fall into this category. The first event took place in 1599 – a year of abnormally heavy rainfall and floods in the lowlands.

The town of Tehuantepec … was suddenly flooded. The buildings and the gates were burst open with violence, and a great body of water inundated the patios, the rooms, the storehouses and cellars. Some walls collapsed and some entire buildings were levelled … the roofs and beams of the houses caved in, and were floating atop the waters in which were mixed dead bodies of lambs, oxen, horses and men.[115]

Independent sources suggest that an event of similar magnitude occurred 122 years later. The event again affected Tehuantepec when 'a storm of strong winds and rains and lasting ten hours from the night of Friday 29 May [1721] till five o'clock in the afternoon on the Saturday [30th]' was recorded.[116] As with the 1599 flood, the River Tehuantepec reacted first, its water level rising so suddenly and violently and with such force that the consequent flood demolished 60 houses, 'all made of adobe and with thatch roofs and in all cases taking all the goods of the residents, leaving neither clothes nor anything with which they might be helped'. Violent winds also wreaked havoc on the local infrastructure. 'The wind blew with such force that it broke apart and uprooted the walls … and lifted the roofs from the houses'. The scale of destruction was, if anything, even more dramatic than the earlier event:

Everything looked as though it had been shipwrecked … wild animals were left in fields, drowned and then afterwards … raised above the trees … the cows and pigs and mules were transported across distances of 22 leagues … dragged along by the flood waters from the town of San Juan de la Harcia right up to Chicapa a little higher up. Some of the grazing stock were drowned in the torrents and great quantities of water and the floods of the rivers and arroyos and others suffered hunger, the pasturelands being totally covered and hidden by a blanket of sand.[117]

Interestingly, the author of the document draws direct comparisons between this event and that of 1599. 'Such an extraordinary event had not been seen of or heard of in past centuries for although tradition knows that in the year past of 1599 the river took away 45 houses it is because they were so close to the river.'[118] Clearly, the devastation caused by the 1599 event was sufficient for it to survive in the social memory of the community, and to represent something of a benchmark against which the impacts of the 1721 event were compared. The suggestion, however, is that such catastrophic events were relatively rare. Moreover, there were some subtle differences between the two events. Attention is drawn, for example, to the fact that the losses following the 1599 event were ascribed in part due to choice of settlement location. The fundamental difference in impact, however, relates more to their respective timings. The event of 1599 took place in September, towards the close of the rainy season when the level of water in the river was high.

Documents indicate that communities living along the river were accustomed to storing up foodstuffs throughout the wet season, should a flood occur as the water levels rose at the end of the season.[119] Thus flooding at this stage in the calendar was a recognized occurrence and had been integrated into agrarian practice. Prior to the event of 1599, therefore, the local populations had been guarding against a possible flood, storing foodstuffs and grains. The event of 1721, in contrast, took place at the beginning of the rainy season when 'the land was dry and needy'.[120] An unexpected flash flood ensued, hitting a consequently ill-prepared population. This was a time when the crops had only just been sown, and when food stores were depleted after the long dry season. Thus, although the event of 1599 was clearly devastating, the economic losses of the 1721 event were correspondingly greater and in fact were estimated to have been in the region of 30,000 pesos. In interpreting the scale of the impact, some consideration must also be given to the greater level of economic and agrarian development of the region by this stage. There had effectively been more development by 1721, and consequently more to lose. Thus there appear to have been more significant economic losses incurred during this event, even though the actual flood may have been of similar magnitude to that of 1599.

Perhaps most interestingly, if somewhat paradoxically, these most damaging and devastating of events, perhaps because of their perceived rarity, stimulated very few major changes in lifestyle or behaviour. Rather, there appears to have been a degree of disaster amnesia or at least an element of deliberate risk taking. The riverside location of many of the most fertile and hence most sought after agricultural plots rendered society in these areas much more vulnerable to flood hazards. Yet it seemed that the perceived productivity gains and the potential for capturing water for irrigation in these areas far outweighed the apparent threat of potentially damaging flooding, not least because in the local social memory of climatic variability in this region, the most devastating floods appear to have been relatively rare phenomena.

Responding Strategically: Climate, Consciousness and Experimentation

The way in which climate changes and weather events are perceived and experienced by people and the knowledge of events in the recallable past determine whether they become inscribed into the memory of those people in the form of oral history, technological adaptation or narrative (Hassan, 2000). These different ways of recounting the past represent central media through which information on past events is gathered and transmitted across generations. In colonial Mexico, social memory of droughts, floods and other weather related phenomena and their impacts conditioned the way in which societies comprehended and also dealt with the problems of uncertainty, risk and preparedness with respect to climate variability and extreme weather events (Koselleck, 1985; McIntosh et al., 2000). Repeated experience of periods of crisis and calamity improved the knowledge of risk among affected communities, increased their awareness of their own vulnerability and thus may have contributed to 'fundamental learning' in this respect (Pfister, in press).

Traditional agrarian societies are often regarded as risk averse, adopting precautionary, sometimes experimental strategies for coping with climatic variability and potential food scarcities (Wilken, 1987), but for the most part based on inter-generational transfer of knowledge grounded in past experience. Purposeful mixed farming, planting a diversity of crops and crop varieties, and the development of non-farm incomes were all important diversification strategies in Mexico, and were adopted at one time or another as proactive responses to climate related problems. Many of these strategies appear to have prevailed and even became more formalized or standardized in the colonial period.

Yet there was also a range of more experimental adaptive strategies. There may have been, as there are in present-day Mexico, attempts to predict seasonal climatic conditions (Eakin, cited in Liverman, 2000: 43). A number of almanacs, for example, refer among other things to predicted growing season conditions, timings, potential weather events and possible crop losses for different parts of Mexico. Almanacs from the 1690s, for example, which as shall be discussed in chapter 6 represented something of a disastrous decade, chart the likely timing of growing and harvest seasons and include daily weather predictions. The Almanac for 1690, for example, forecast the weather for each month of the following year. Sunday 1 May was predicted to be 'damp and cloudy with very cold winds', while the full moon on the 22nd of the month, a Monday, would bring rains if it was windy, but they would not necessarily be torrential.[121]

There is also evidence to indicate that there was some agricultural experimentation in order to avert climate related harvest losses. The most common time for sowing wheat in Guanajuato, for example, was in the spring in order to capitalize on the summer rains when the crop was maturing. Harvest losses due to drought, or more commonly the late arrival of the rainy season, as illustrated, were not uncommon and were regularly recorded in the archives. Some agricultural communities in the region responded by attempting to grow a second 'experimental' wheat crop in the late summer and autumn in the hope that late (or even winter) rains might arrive to ripen it. The seedlings were known as trigo aventurero and several references indicate that some farmers resorted to this strategy during dry years or years when the rainy season was noticeably delayed.[122]

There was also, however, a degree of experimentation when it came to grazing pasture. Shortage of sufficient pastureland and a lack of water during particularly severe drought periods regularly destabilized the livestock industry, particularly in the north of the country, with knock-on consequences for other sectors of the economy. The drought reported in January 1758 contributed to a shortage of grazing pasture in Chihuahua.[123] There were calls for 'common pasturage' by some people and lands normally considered unsuitable for grazing began to be exploited in order to avoid crisis in the livestock industry. There are also references to cattle ranchers making use of salt flats and swamps in the area for grazing in January of 1758. Notwithstanding such requests and adaptations, 'total ruin among the livestock' was recorded.[124] Twenty years earlier, in 1738, a mysterious illness had swept through the cattle populations of the region. It was argued that the illness related to the cattle grazing a poisonous grass referred to as Garuansilla, though others suggested that cattle had grazed this species at other times of pasture failure and had not been affected.[125] That the cattle might have been forced to graze more noxious pasture, however, is a real possibility given the scale of the drought recorded in other documents for this period and the likelihood that, as with later droughts, pastureland was consequently in short supply. Elsewhere, however, there was more success in finding alternative grazing pasture. During the drought of 1785–6, the growth of a grass species in the district of Zamora, northwest Michoacán, capable of surviving drought, provided cattle fodder where all other plant species failed. In consequence, there was said to be 'more milk than water' in this location.[126]

Agrarian experimentation also took on a more long-term, strategic and scientific form. Periodic crop blights contributed to widespread agrarian crises and food scarcities in colonial Mexico on a number of occasions. Wheat crops across the country were repeatedly if selectively destroyed by a crop blight referred to as Chahuistle.[127] Gay, for example, highlights the impacts of Chahuistle in Oaxaca in the 1690s, a time, as shall be demonstrated in chapter 6, of unusual phenomena. 'The most notable of these circumstances', Gay suggests, 'was chahuistle, a wheat blight … which caused great harm to sown crops across the whole of the country' that year (Gay, 1950: 385). In fact, wheat harvests in the Etla Valley of Oaxaca would be repeatedly affected by chahuistle between 1696 and 1714.[128] This extended blight led to the abandonment of (wheat) flour mills in the region, a number of which, by 1714, were said to be in ruins. Wheat crops in the Bajío were similarly affected on a number of occasions, most severely in 1700, 1706, 1711, 1718, 1735 and 1746, stimulating shortages across the region.[129] There were, however, some attempts to develop a remedy for this persistent scourge. A document from Léon in Guanajuato, dated 1782, details a 'secret recipe' for ensuring resilience against crop (specifically wheat) loss to chahuistle and which stated that it not only 'prevents chahuistle', but also 'increases the amount of harvest in proportion to the quality of the land'. The document describes precisely what the preventative measures involved:

> Firstly, in a tin or canister fit a third of wheat or one carga and enough water to moisten the seed, with care taken to ensure that the tin is emptied out before-hand, and then bring the tin to the boil in a suitable container with four ounces of white arsenic if you have used one carga of wheat or if you have used a third, use two ounces of arsenic, it is better if this is in powder form; and leave the boiling water standing and mix in the dust; immediately take it from the light and with the same boiling mixture immediately add it to the wheat in container mixing it together very quickly with a stick ... after about half a minute drain the water off and put the mixture on a petate ... and sow it instantly, because if you keep it for another day the whole procedure will be wasted.

A second method using sulphur in place of arsenic, 'which is the one used in Europe', was also recommended.[130] Both recipes had been devised by a Don Pedro Aspe, who was of French origin but who 'noted that in most years there had been excessive prices for wheat because of the ruin of the chahuistle con-tagion'.[131] Dissemination of this hitherto 'secret' information represents an important step in the agrarian history of the country and a key development in social adaptation to climate variability in this central region. It also indi-cates that there was some longer-term research and development and, more-over, international dialogue in terms of reducing agrarian vulnerability.

The upheaval, economic costs and, on occasion, devastation and loss of life associated with climatic variability, periods of unusual weather or weather related events thus prompted a variety of remedial, mitigating and experimental actions, coping strategies and adaptations to reduce vulnera-bility. While recognition of common vulnerabilities might have facilitated a degree of collaboration and co-operation between different cross-sections of society, however, climatic variability, extreme events, and the crises that sometimes followed could also serve to at once reveal and highlight tensions between social groups. Indeed, an awareness of seasonal variations in water, or knowledge of the likelihood and implications of drought or flooding in particular regions at specific points in time, is equally manifest in the legal proceedings over water rights, access and management in colonial Mexico. Archival references indicate that prolonged or successive droughts in par-ticular, by leading to harvest failure and famine, disease and death, loss of economic livelihood, land abandonment, and out-migration, may have also played a role in triggering social unrest and physical violence. Chapter 5 explores the way in which climatic variability and changes in water availabil-ity might have contributed to legal and, in some instances, physical conflict in colonial Mexico.

Chapter Five

Dearth, Deluge and Disputes: Negotiating and Litigating Water and Climate in Colonial Mexico

Introduction

There is a vast literature on the causes of social unrest, rebellion and insurgency in Mexico. This has illustrated the importance of considering cultural and structural factors, and of incorporating analyses of the actions, powers and adaptations of elites, states and the urban and agrarian masses. The significance of environmental change and also climatic parameters as 'triggers' or stimuli of social unrest in Mexico has also been highlighted (Liverman, 1990). Florescano et al. (1995) and Tutino (1986), for example, have argued that social unrest in northern and central Mexico increased due to high corn prices following drought induced famine across the country in the 1780s, and again during a period of harvest crisis just prior to the wars of independence.

Yet there has been a tendency to view the struggles of the later colonial and independence periods almost as isolated historical episodes of unrest (Hamnett, 1997). Although undoubtedly among the most turbulent times in Mexican history, the social, class and political struggles of this period represented the culmination of a much longer-term record of dissent which, Hamnett (2002) suggests, spans the immediate post-Conquest period through to the Liberal Reform Movement and early stages of modernization in Mexico. One could even position the late colonial and independence movement within a longer time frame that incorporates periods of pre-Columbian unrest. Pre-Hispanic peasant uprisings in central Mexico in the fourteenth century, for instance, highlight a class consciousness that would be echoed in the drive for independence (Katz, 1988). One must also be cognizant of the very many different spatial scales and contexts of unrest and dissension, within a variety of rural and urban settings, and at the individual, local community, institutional or regional scale throughout Mexican history.

Indeed, adopting Tilly's (1996) terminology, a broad 'repertoire of contention' can be identified in colonial Mexico. Firstly, there were local and most commonly indigenous rebellions aimed chiefly at eliminating particular grievances with the colonial administration. Usually of relatively limited duration, these largely came about as a result of administrative abuses, unreasonable tribute demands (especially during times of dearth and food scarcity when such demands became impossible to fulfil), unfair labour practices and conflicts in religious or cultural interests (Hamnett, 2002: 74). While there is a definite rurality to much of this type of social unrest recorded in colonial Mexico, urban grain riots might also be considered within this general schema, particularly when one considers that there was considerable inter-digitation between rural and urban systems in the colonial political economy.

There were also many different forms of what Scott has referred to as 'everyday' or 'Brechtian' forms of resistance in ongoing processes of class struggle that included 'foot dragging, feigned ignorance, slander, arson, sabotage' (Scott, 1976: 28–29). This type of dissent, Katz argues, can be largely associated with communities, especially those in those regions that retained a strong indigenous character following conquest and throughout the colonial period, such as can be found in the southern regions of the country.

Perhaps the most common and persistent manifestation of social tension was the lawsuit – a much more local affair. Conditions on the land and problems of sharing resources, specifically land and water, between competing users frequently led to localized legal infractions between all sections of society. Indeed, some native communities in the central and southern regions of the country gained a reputation for being especially litigious and for flooding colonial courts with complaints of abuses in local (village) government (Serulnikov, 1996: 189). The colonial legal records for these regions in particular reveal considerable tension at the local level over land and water resources throughout the entire colonial period. By far the majority of cases were indeed raised by communities against local landowners or landowning institutions, commonly the Church, but disputes between individual, often neighbouring, landowners, or between communities were not uncommon. Although not always as noteworthy as riots, revolts and rebellions, these forms of resistance were far from trivial. Literally thousands of such documents have been filed for the colonial period alone, some spanning generations, representing a background but persistent level of dissatisfaction.

An example of more widespread and more violent regional level resistance, in contrast, and which required governmental and military intervention, might be associated with the unconquered and (as already illustrated) bellicose and largely nomadic or semi-nomadic peoples of the northern frontier of settlement, and prevailed in various forms and levels of intensity from the time of European contact throughout the colonial period through to independence.

The purpose of this chapter is to explore some of the different elements of this longstanding and varied record of unrest in colonial Mexico, and to highlight the degree to which climatic variability, changes in water availability and security, and a general climatic and environmental consciousness might have contributed to different forms of dissension at a range of spatial and temporal scales. The aim is to demonstrate how legal disputes, at the individual, community or institutional level, protracted and violent regional-level resistance as well as short-lived and localized uprisings and riots and rebellion, might have represented reactions and responses to real or perceived social and biophysical vulnerability to changes in water availability and accessibility, climate variability and extreme weather events.

Water and Local 'Everyday Conflicts' in the Country and City

Perhaps the most significant and long-term agricultural adaptation to climate variability, periodic drought and the constant threat to water security in Mexico was, and is, the use of irrigation. Permanent and ephemeral watercourses, rivers and arroyos had long been exploited by pre-Hispanic populations for irrigating *milpas* (maize plots)[1] and a variety of crops, including garlic and beans.[2] After Spanish contact, however, there was a need to ensure possession of and access to water supplies, not least because of the seasonal requirements of the introduced Mediterranean wheat varieties, which were accustomed to wetter winters and summer drought.[3] There was considerable expansion in the area of land under irrigation, particularly in the Bajío, which as noted became an important cereal growing area. Access to water sources was also fundamental to the livestock economy and for mineral processing, and was obviously essential for domestic purposes, while some water bodies were prized for their fish stocks. Water thus became very much a contested resource between different users and stakeholders.

Legislation concerning water management, ownership and distribution appears to have been only partially effective in resolving distribution difficulties. Even where formal mercedes existed, grants of water were often enjoyed only 'on paper' (Lipsett-Rivera, 1993). Illegal abstraction and private appropriation of water for irrigation purposes, and use of water that had not been subject to the formal granting process, caused considerable problems for both rural and urban populations. Perhaps not surprisingly, therefore, there are literally hundreds of recorded lawsuits in the colonial legal documentation, citing instances of water monopolization, deprivation, usurpation and over-abstraction. Local contention over access would escalate during times of actual drought, or even feared or predicted water stress.

The following sections explore a number of examples of local water pleitos: those raised by communities against individuals or the Church, and pleitos between individual, often neighbouring, landowners and hacendados. The way in which drought may have stimulated or been implicated in such disputes will be highlighted and the consequences of water sharing strategies designed to overcome problems of unequal water distribution will be discussed. Attention then shifts to local pleitos over flooding and the degree to which water management infrastructure was recognized and highlighted as exacerbating flood risk.

Representing the community: water pleitos and subaltern voices

Crown legislation afforded only theoretical protection for Indio communities. Despite a variety of legal instruments governing water access and use in colonial Mexico, indigenous communities could and frequently did find themselves without sufficient access to water. Moreover, even where there was a good deal of indigenous land retention, for example, in the Valley of Oaxaca, where some communities managed to retain control over considerable properties, access to water that ran within these lands was not necessarily guaranteed. As suggested in chapter 3, the majority of mercedes awarded for water, or granted for land with associated waters in colonial Oaxaca, for example, during the first century of colonial rule were awarded to Spaniards/Creoles and the Church.[4] Grants of water to Indios were far less common and, where they exist, appear to have invariably been awarded on a community or pueblo basis.[5] Perhaps for this reason there are very many documents filed throughout the colonial period that request protection for communities whose access to water was threatened.[6] Community representatives were often responsible for bringing the cases to court. In this way, litigation documentation highlights the consciousness among the indigenous classes of a deteriorating or threatened status (Hamnett, 2002: 3).

Representatives of indigenous communities across Mexico were in fact very active in defending their rights to water, though most commonly through collective effort (Lipsett-Rivera, 1999b: 17). When written documents confirming rights of access did not exist, were fragmentary or went missing, the general practice was to obtain oral testimony from elder members of the community to establish past usage and rights of access. Pre-Conquest water possession was usually respected as a source of title and many of the pleitos raised by indigenous communities make this case. As Lipsett-Rivera (1999b: 25) suggests for Puebla, physical evidence of old water works could also be used, though rarely, to prove antiquity of usage,[7] and in some instances, she suggests indigenous communities also resorted to forged documentation in an attempt to secure water rights. Competition for water frequently led to legal conflict and, in some rare instances, physical violence. To some extent, circumstances such as season and location dictated the nature and form of the contestation, and whether it was passive or active (Lipsett-Rivera, 1999b: 105).

Lawsuits between communities and neighbouring estate owners were the most frequently filed of all water related pleitos. This is perhaps to be expected, since irrigation for large-scale cultivation often entailed the diversion of water from location to location[8] and in some cases this effectively meant the removal or withdrawal of water upon which local communities had been reliant, chiefly for maize cultivation, for generations.[9] In some situations the merced system was used to formally ratify water sharing arrangements. In 1621, for example, the Real Audiencia awarded a merced to Don Gaspar Juarez in Teotitlán del Camino, Oaxaca, to use water from the river in the town for his hacienda and wheat mill. It seems that the indigenous community already made use of half the available water, storing the water in a small reservoir and raising it to the reservoir via a small ditch which they had constructed themselves.[10]

There are many more cases, however, which deal with complaints about illegal diversion of water sources, in some cases from the earliest stages of European colonization. In Celaya in 1573, for instance, complaints were made about Don Juan de Yllanes, who was diverting more than a half of the water from the River Laja for the benefit of his lands and those of Ponce de Léon, so depriving the local residents of the town of much needed irrigation water for their own plots. The commissioner responsible for land administration in the region resolved to confiscate some of Yllanes' land and decreed that neither culprit should be permitted to irrigate their fields until the residents of the town had first irrigated theirs.[11] Several later lawsuits describe such acts as instances of water theft. A legal document filed in 1634, for example, details a dispute again between the residents of Celaya, a number of landowners and the local council and a group of hacendados who were charged with 'stealing' water from the river to irrigate their wheat fields. Several witnesses provide evidence to suggest that the town's residents had as a result suffered from water scarcity for a period of 25 days. According to a number of other witnesses 'the river no longer contained water' because of the actions of upstream hacendados.[12]

Episodes of drought might have increased the likelihood of litigation over water in some instances. The severe drought that affected the whole of central Mexico between 1590 and 1591[13] stimulated pleas for assistance from local communities, in one instance for water for their beasts of burden.[14]

This appears to have been a fairly widespread and severe drought, causing, among other manifestations, the almost complete desiccation of Lake Cuitzeo just over the border of Guanajuato in Michoacán.[15] It was in this context that Don Luis de Velasco Avosalo Perez de Bocanegra, Alcade Mayor of Celaya, was asked to support the Indio population of the town in their claims that they had experienced 'common damages and disputes' over water, and specifically the branch of the Rio Grande known as Toluca, because of the actions of a number of Spaniards who had agricultural lands and ranches in the area.[16]

Similarly, the drought of 1695–6 which resulted in food scarcity across much of central Mexico might have also contributed to a number of water disputes. A document from 1696 details a dispute between the residents of the town of San Juan Bautista de Apaseo and a Juan Garcia de Alarcon over access to water. The residents had a concession to 'half of the water to irrigate their fields', but this was apparently being threatened by the actions of their neighbour Alarcon, who was apparently 'irrigating a great part of his lands in a year so calamitous as the one currently being experienced'.[17] Extravagant use of water during a year of scarcity here provided the local residents with cause to complain.

Representatives of missions were regularly accused by indigenous groups of monopolizing water during periods of drought. It was during a well-documented period of drought and food scarcity, for example, that an individual named simply as Juan, Governor of the Indio town of San Geronimo near San Felipe el Real de Chihuahua, filed a case against the mission known as Nombre de Dios. According to the pleito, which was dated 4 May 1729, following a number of years of drought and food scarcity across the region, the mission was being charged with monopolizing and usurping the water supply. 'The municipal legislation,' stated Juan, simply 'did not favour the Indios' of the town.[18]

Many of the testimonies in water related pleitos explicitly refer to seasonal variations in precipitation or actual periods of drought as being central to the problem of water availability and distribution. One such case from 1690 concerns the Indios of the pueblo of Teotitlán, in the Tlacolula Valley of Oaxaca and the Jesuit Colegio[19] of Santa Ana over the shared use of the waters of the Salado River.[20] Representatives of the community indicated that they had long been 'in peaceful possession' of water belonging to the town, but because of the actions of the local Colegio, the community had apparently been left in desperate need of water for their crops, gardens and fruit trees. According to the community, the problems dated back to 20 September 1621, almost seventy years previously, when a merced had been granted to the Colegio to use half of the water from the shared source to run a sugar mill. However, as the representatives of the community pointed out, when the grant was made ... it was ... during the rainy season, a time when there abounded more water in all parts, now ... there is a diminished amount in the dry season ... there is not enough for all, nor sufficient for the said town ... the land is dry and sandy and extends where the river ran and it is wrong that it is possible to have a sugar mill running when the town lacks necessary water.

Seasonal variations in water supply had, it was argued, clearly been ignored when the merced was approved. Moreover, the community argued that that the mill's needs had changed since the merced was originally awarded. 'When the said merced was awarded the [lands of the] mill was very small and they sowed only a small amount of land for which only a small amount of water was sufficient.' They go on to suggest that an increase in the land sown, and hence the amount of sugar cane being processed, had necessitated an increase in the amount of water required to run the mill, stating that 'the sugar cane lands have now been sown not only with all the water of the town but with much more because they have procured three springs or sources of water located within the limits of the mill's lands'. According to the community, the result was an apparent monopoly of water sources in the area such that for the community of Teotitlán there was 'not even any drinking water' available.[21] From this testimony it appears that water scarcity was a direct consequence of a progressive monopolization of water supplies by the Colegio, the impacts of which appear to have been disproportionately felt by the local community in the dry season when the water supply was naturally limited. One must indeed consider the nature of the watercourse itself. The water level of the Rio Salado fluctuates throughout the different seasons of the year and often desiccates completely during the dry season (Taylor, 1978: 11). That the water scarcity experienced relates to the natural condition of the river at this time of year should thus be borne in mind. A series of testigos (allegedly impartial witnesses), however, forward a fresh argument that suggests that the problem that the community was facing was a clear function of climate events. One Sebastian Lazo Sastre, for example, noted that 'in forty years of residence in the town I have seen the rains between the month of September until November. In other years there have been rainstorms, but in the present year there has been a very rigorous drought.' Another witness, Antonio Garcia, further supports this view, but declares that he has lived in the town 'for three years during the course of which time there has been drought'. On this basis it appears that the community's case for water deprivation possibly had less to do with the Colegio's monopolization of the shared water source than a period of 'rigorous' drought and perhaps even consecutive droughts over a three year period.[22] It was not unusual for one or other party to pay supposedly impartial witnesses to support them in such pleitos. However, as shall be demonstrated in the following chapter, evidence from numerous sources suggests that the period between 1690 and 1692 was particularly calamitous across Mexico. Thus, in this case, the impartial testigos may have been just that, providing much needed objective evidence.

Some pleitos filed by communities against individual property owners charged with the monopolization of water spanned many years and even generations and reflect a longstanding competition for water sources in a particular location rather than any response to specific periods of water shortage. One long-running dispute between the indigenous community of San Miguel and the Spanish residents, in what seems to be a separate Spanish enclave in Guanajuato, focuses on the use and misuse of a spring of water known as El Chorillo (meaning 'the constant stream'). The lawsuit spans close to a century. The first legal disputes between users of the water source were recorded in 1655, but appear to have continued to 1745. According to the papers filed in 1745, the Indios of the town had suggested that 'the water which originates within the town has been demarcated half to the town and the other half to the Villa de Españoles' for the local administration there. The problem had arisen 'because it was found that almost all the water is channelled to the Villa, leaving such a small amount flowing, that not only do they [the Indios] suffer the loss of their sown crops but they lack the water provision necessary for the people'. The indigenous population of the town had thus been left in a 'miserable' state and so requested a more equal division of the water. A reconnaissance trip undertaken by a local official during the consideration of the case indicated that not all houses and residents had the same allocations of water. Indeed, this lack of legal division was causing 'many violations and leaving some with nothing and letting others take as much as they want'. It was thus finally agreed that a formal division needed to be drawn up to avoid further lawsuits and infractions. Furthermore, this division would include provision of water for other community needs, including 'a well and wash places ... where they can gather to bathe'.[23]

Another pleito charts a 35-year dispute between the community of Etla, in Oaxaca, and local mill owner Luis Rodrigues over his apparent monopolization of river water,[24] while a pleito between the Convento de Santo Domingo and the Antequera city council spanned over 154 years.[25] Yet it should be noted that water sharing agreements were also frequently reached in colonial Oaxaca, many as part of the rental arrangements drawn up between cacicazgos, communities and individual landowners. The agreement recorded and confirmed in Etla, Oaxaca in 1733 between the Indios of the town of Nuestra Señora de Guadelupe and local landowner Don Pedro Garcia del Barrio was typical of many. It seems the community was prepared to give Don Pedro access to the water that ran through the community's lands via an aqueduct, on condition that he pay them 15 pesos each year for the privilege. The final declaration adds that this arrangement was made on the basis that water was not to be used all year round or in 'large quantities', but just during periods when irrigation water was required.[26]

Struggles over cultural norms and values might well be evident in such water conflicts between indigenous communities and local landowners. Indeed, water may have served as a medium that effectively framed racial and class relations in colonial Mexico. Dissent over water distribution, however, was not a solely dichotomous contest between indigenous and Spanish colonial authorities. Pleitos between different indigenous communities were also filed, although these lawsuits tend to be associated with longstanding disagreements and were often related to access to land as well as water. Perhaps significantly, these pleitos refer less to drought induced water problems and to more general problems of water scarcity due to human action.[27]

In some periods, however, disputes between neighbouring pueblos had the potential to become rather inflammatory. One such case occurred close to the start of the summer rainy period of 1815 between the communities of the pueblos of San Lorenzo and San Felipe in Huitzo, Oaxaca over the maintenance of a water ditch and use of the waters of the River Xalapillo. According to the people of the town of San Lorenzo, they had held rights to the use of a portion of the water 'for many years and not less than a century' to irrigate their wheat fields. The dispute focused on a water intake or channel which both communities were responsible for clearing of silt and debris. One Indio witness was 40-year-old Domingo Diaz, who lived in the town of San Lorenzo. Diaz said that the people of his town had agreed to meet with those of San Felipe to clear the channel 'as they were obliged to do every year at this time in order to drain the waters of the river to their fields which they had prepared for sowing wheat'. Having arrived at the predetermined meeting place, they started work. A short while later, however, the people of San Felipe arrived, but refused to participate and threatened to destroy the channel, arguing that it was theirs and it fell within their lands. Violence had broken out between the communities, involving 25–30 people, by the time the Alcade and Regidores arrived to adjudicate According to another witness, the throng 'began to throw stones and rocks, and others took up knives and chased people into their houses'. The surgeon who treated those injured in the fray described the wounds and contusions as 'very serious'. There was particular concern among the people of San Lorenzo for a swift resolution. They claimed to have land and water rights which the 'people of San Lorenzo knew about' and that it was 'urgent' that the difficulties were resolved because their fields were prepared and ready for irrigation.[28] Unfortunately, it is not clear how the communities' differences were resolved. However, the timing of the dispute and confrontation may well be significant. As shall be illustrated in the following chapter, this was a time of widespread grievance and a time when drought, food scarcities, and epidemic disease in the previous years had highlighted deep structural problems, inequality and differential social vulnerability within society in all three regions. This was also a time when there was considerable unrest across the country as the drives towards Independence were gaining momentum. Thus, while in most cases actual physical unrest over water sharing and distribution was rare, tensions over this most vital of resources might have been correspondingly greater at this time, accounting for this violent infraction.

Manipulating the system? Drought speculation and water conflict between landowners

The control of water was seminal to all sections of society (Berry, 2000). Thus, while many of the litigation documents deal with disputes between Indio communities and Spanish or Creole landowners over access to territory, water and other natural resources, conflicts between hacienda or individual landowners were not uncommon (Endfield and O'Hara, 1997; Endfield, 1998). Many times, as shall be shown in the following examples, actual drought, fear of drought, seasonal variations in water availability and knowledge and experience of past water scarcities due to climate variability may have played a pivotal role in driving some of these cases.

In each of the three case study areas, many of the most prosperous landholdings were located along rivers, with owners often constructing dykes and ditches as well as small mud earth dams (presas) to divert river water for irrigation and/or domestic use. Depending on their location, the amount of water abstracted by one user could influence or be influenced by the uptake of water by other users. Water disputes were thus especially common where neighbouring haciendas shared the same watercourse. Other lawsuits chart the illegal damming of rivers and water channels and the implications that these actions had for downstream users. By directly reducing the amount of water available for irrigation as well as other uses, however, dry season conditions and, in some situations, periods of drought might have aggravated the degree of tension over shared water sources, leading to intense legal contests. In 1738, for example, a pleito between the owners of the haciendas of Valdeflores and Buenavista in the southern Zimatlán Valley, Oaxaca, over the use of the water of the River Atoyac appears to have been driven by a fear of impending drought. Don Juan Antonio Jimino de Bohorques, resident of Antequera and owner of Valdeflores, charged Juan de Antelo, dueño of Buenavista, with the 'violent and thoughtless' over-abstraction of water from the River Atoyac which passed through Bohorques' landholding,

> and with it now, at the beginning of this month of December, he is irrigating wheat or maize without a thought to the grave damage that this is doing not only to my [Bohorques'] livestock but also all the towns and haciendas of the valley ... and as my hacienda is situated 5 leagues lower down, so little [water] arrives that it is said I have none.

Bohorques feared he would face even greater problems in March and April, the driest months of the year, and predicted that 'a year of drought' lay ahead. His desperate plea to Antelo did not help matters. 'Although I went personally to inform him of the damage ... asking him to leave some water, he responded to me negatively, telling me that it was for public use.' In his defence, Antelo suggested he was conforming to the legal distribution of water and was mindful of the 'general scarcity in water from the aforementioned river'. He dismissed the accusations against him as 'false', arguing instead that haciendas further upstream had played a greater role in reducing the water supplies. Drought speculation, rather than drought per se, may have been used as supporting evidence in legal proceedings against Antelo in this case.

By denouncing another's lack of title to water, a landowner could effectively secure a grant of water for himself.[29] Individual landowners, therefore, would often exploit this facility in water disputes, especially where their properties shared a water source. Yet in these cases too, knowledge of drought may have been used opportunistically. One such case was filed in 1797 between Don Francisco Garcoa, hacendado of the estate known as Cinco Señores, and Don Florentino, hacendado of De Cañas, both properties being located in the vicinity of Antequera. On this occasion, actual experience of drought seems to have been a key factor driving the case. Both parties were dependent upon the waters of the Rio Xalatlaco to irrigate their lands, which were located downstream and upstream, respectively. Don Francisco paid a rent for the use of any water remaining in the river by the time it had passed through his neighbour's lands, stating that the natural course of the river ran through his landholding and hence it was his right to extract its water. This arrangement, however, was periodically compromised. 'In times of drought', Francisco claimed, 'all the water is consumed before it arrives into my land and by reason of this, much of the irrigated crop has been lost.' He continued: 'one such year was the year just past ... the cause being that the hacienda de Cañas has already claimed it [the water].' Suffering a loss of water for irrigation due to apparent drought conditions in the previous year, and being at the mercy of his neighbour for any water that did reach his land, Don Francisco demanded that his rights to this water were honoured by the city magistrates.[30]

In some instances scarcity of water during times of drought appears to have exacerbated other ongoing disputes between neighbouring landowners. One case in point was that between Francisco Javier Maldonado Ovalle, resident and Regidor of Oaxaca and owner of the hacienda of Nuestra Señora de Buenavista, against Joseph de Zabaleta, who rented a ranch within the lands belonging to the people of the town of San Pedro Apostol. In 1728, Maldonado was taking 'civil and criminal action' against Zabaleta for encouraging his servants to kill and remove his cattle from the hacienda. Zabaleta had already fallen out of favour with the community for keeping sheep and goats on the land which were incompatible with the cattle. His actions were understandably regarded as intolerable, but all the more 'insupportable' and indeed 'fatal because of the lack of rains'. By introducing more livestock to the area, water sources were even more stretched than usual. In his defence, however, Zabaleta claims that, because he had kept up his rental payments, neither his ranch nor any of his property could be forcibly reclaimed.[31]

Other pleitos chart the more desperate actions of individuals during apparent periods of drought induced water scarcity. A document from 1638, for example, deals with a case of water theft for irrigation purposes in Oaxaca. Investigation into the case revealed that the servants of a local government official, Don Joseph Delgado of Antequera, had, upon his instruction, 'stolen' water from the city's main water supply to irrigate wheat and sugar cane on their employer's hacienda. The accused claimed that his actions were a necessary and 'desperate' response to general water scarcity at the time.[32]

Sharing the essence of life: rationing and rationalizing water use

Disputes over the shared use of water would often result in the introduction of the royal guideline of 'reparto de aguas por dias como le parece', which was, at least theoretically, an egalitarian system of diurnal/nocturnal water sharing between users.[33] Indeed, as seen in the example from the Valle de Santiago in the previous chapter, where there were many different users of water in a single irrigation system, water rights might also be allocated by tandas (Lipsett-Rivera, 1999b: 20). One such case was recorded in Huajolotitla, Oaxaca in 1713. The Viceroy, Don Fernando de Lencastre Noreña y Silva, Duke of Linares, instructed the water judge to award free use of water from the River Zalapilla to Captain Don Pedro Gonzales de Mier for a period of 'fifteen days with associated nights.'[34] Such essentially temporal divisions, however, could be quite complicated and provided opportunities for legal challenges. The people of the barrios of Etla appear to have been particularly litigious in this respect. One dispute over shared use of the Atoyac River between the neighbouring communities of Soledad de Etla, Nazareno and Guadeloupe Etla lasted over eighty years and resulted in the imposition of the water sharing guidelines, first in 1703 and then again in 1774 when a local magistrate finally apportioned fourteen days of irrigation to Soledad, nine to Guadeloupe and seven to Nazareno, based on ratios of landownership nearest the water source.[35]

Yet even where such arrangements had been legally agreed, it was not unusual for one or other party to flaunt the legislation. Indeed, in some instances, sharing of water based on an alternate day/night cycle seems to have positively prolonged disputes.[36] One document from 1718, for instance, deals with a dispute between Don Nicholas Molino, resident of Celaya, and Antonia Guitron. Molina was paying rent to Guitron for two days of water to irrigate his fields. Guitron, however, was apparently 'impeding the flow of water' to ensure his own water needs were first satisfied.[37] Similarly, the granting of diurnal or nocturnal water rights often became complicated when people attempted to transfer allocations from person to person. A document from 1707, for instance, described the difficulties surrounding a six day water allocation awarded to Don Joseph Comargo, resident of Celaya, which he used to irrigate his land in return for maintenance of the channels and ditches used to conduct water to the city. Being unwell, Comargo had asked that the contract he had with the council should be revoked, so freeing him from the task he was charged with. His wish was that the licence be passed to Don Nicolas Molino, though this transfer involved quite protracted negotiations.[38]

Research elsewhere in colonial central America has identified how unrest over access to land and other resources increased throughout the colonial period (Lipsett-Rivera, 1990a; Webre, 1990). Lipsett-Rivera (1990a, 1999b), for example, has demonstrated how disputes over water in Puebla escalated during the eighteenth century. In part this related to socioeconomic developments and land use changes. Deforestation, to provide timber for construction and to clear land for sowing, increased run-off, reduced the retentive capacity of the soil and lowered the water table. Expanding populations throughout this period also effectively increased demands on available water sources, while structural differences in land tenure necessarily led to differential access to this most essential of resources. Similar changes may have taken place in other regions of Mexico. In respect of this trend in demographic expansion and competition for natural resources, any analysis of the relationship between water conflicts and drought has been dismissed by some as 'pointless' (Ouweneel, 1996: 93).

Certainly, a considerable number of the water pleitos considered here relate more to water deprivation than they do to drought per se and by no

means all references to drought in such documents were necessarily legitimate. Indeed, there may have been an element of opportunism in water
litigation and an awareness of the propensity for drought may have been
employed by litigants as a legal 'tool' to reinforce or refute claims of water
shortage, monopolization, deprivation and/or restitution. This seems to
have particularly pertained among the more litigious and opportunistic
communities in the Valley de Oaxaca.

Nevertheless, even if people were capitalizing on the general awareness of
such weather events in order to support claims for land and water rights or
to petition for use of water sources for particular purposes, there are clear
links that can be drawn between drought and legal conflict over water.
Greater consciousness of the propensity for drought, and hence an improved
awareness of vulnerability to fluctuations from normal seasonal rainfall, may
have been metered out – and indeed are manifest in the many and varied
lawsuits over water access and shared use (Endfield et al., 2004a). Litigation
documentation in this way provides a medium through which to investigate
how experience or knowledge of the impacts of drought events may have
been woven into the cultural fabric of legal discourse and practice.

Water management, floods and legal conflict

Of course, it was not only water shortages and drought, perceived, imagined
or predicted, that posed a threat to livelihoods. Flooding represented a
recurrent problem for some communities in colonial Mexico and, as seen in
the previous chapter, particularly for those in the Bajío. The number of
recorded flood events in the region appears to escalate noticeably in the
second half of the eighteenth century. By this stage, a complex myriad of
water diversion channels, dams and reservoirs had been constructed, largely
for irrigation purposes in this, the breadbasket of the Mexican economy, to
ensure crop growth during the winter dry season and even during short-
lived droughts. Many lawsuits were filed detailing cases of water diversion,
storage, monopolization, deprivation, usurpation, dam breaches and over-
abstraction associated with this water management infrastructure.[39] The
majority of cases were raised by individuals against other, often neighbouring, landowners, though once again, community pleitos against hacendados
were also common.

The water management infrastructure, its neglect or misuse forms the
focus of many of these pleitos. People in Guanajuato, for example, were
certainly acutely aware of the way in which water management could exacerbate flood risk. Descriptions of the flood which engulfed Guanajuato in
1770, for example, recorded as part of a survey of damages conducted by a
local administrator, indicate that its origins were initially thought to lie in

heavy and unusual rainfall. 'According to word of mouth', suggests our informant, 'there was a very heavy rainstorm in the hills and in the city, causing a great flood in the arroyo or small channel which crosses many parts of the city.' A reconnaissance trip, however, revealed that there were other exacerbating factors. 'I went out to look at the whole city and its surroundings to ascertain the damage … the cause of the ruin lay in various dams and reservoirs made in the river channel, the various pillars with foundations belonging to the haciendas with which the hacendados procure the commodities of the said arroyo, trapping the waters.'[40] The description indicates that water storage systems in the city had served to exacerbate flood risk; more specifically, the collapse of a dam seems to have been the main cause of this particular flood event. The 'very heavy rainstorm', however, may have been influential in triggering the breach.

A number of similar instances of inundation may have resulted from or stimulated legal conflicts or aggravated existing tensions between different sectors of the community. A document recorded in Celaya on 26 June 1791 between the Mayor of the indigenous residents of the city and a number of local hacenderos illustrates this well. The Mayor, who filed the case, suggests that 'there has been seen twice an overflowing of flood waters which afterwards have caused ruin and damage to the houses and greater damage for the general public', adding that 'the Laja had destroyed our maize plots and the rest of the sown fields, indications are that the floods have completely destroyed our houses and we have been put in imminent danger.' The blame, it seems rested with the local hacendados or estate owners who had changed the natural course of the river with dams, 'causing damage in times of floods'. Direct links were thus drawn by the indigenous community between the imposition of dams and flood risk, for, as the Mayor continues to suggest, 'the water has only to tip slightly over the top of the dam … and a large stretch of territory on one and other side of the land can be effectively drowned in the floods'. There was thus a clear awareness of the environmental, social and economic risks posed by water diversion and storage in the area, which seems to have been based on past experience of flooding.

Seeking some level of intervention from the local council, the Mayor refers back to the level of action taken following a flood event which took place on 7 August the previous year and which he described as 'one of the greatest floods that there has been in this city'. The scale of the inundation and the damages inflicted upon his community were such that the local council had decided that 'all dams should be drained to prevent the floods which have caused local residents to suffer with notable damage to the health and their interests' – a course of action that they felt was justified given that one of the dams had apparently been 'constructed furtively and almost with some degree of despotism'.

Other factors might have exacerbated flood risk at this time, of course. The year had been unusually wet and the river itself was described as 'tempestuous and turbulent'. Flooding had affected communities elsewhere, particularly in the highlands of Michoacán, at this time and roads and walkways across the region needed extensive repairs following the damage caused by the rainstorms.[41] The devastation and disruption resulting from the flooding had additional repercussions. 'Plagues and pestilence' were recorded across central and western central Mexico following the heavy rains,[42] and there was concern among councillors that stagnating pools of standing water were contributing to disease among the communities, 'causing notable damage to their health and interests'. Though it is difficult to draw direct links between the emergence of the epidemic and unusually heavy rainfall, flooding and the prevalence of standing water will most certainly have been conducive to disease diffusion. Thus it was perceived that water storage behind the dams had not only exacerbated the risk of flooding, but that this in turn had serious environmental health implications.[43]

The council's decision to dismantle sections of the water system in the area was nevertheless controversial. Several lawsuits illustrate the degree of unrest voiced by local landowners affected by this course of action. One dispute between Captain Crespo, hacendado of Molino Chico in Celaya, and the council enlisted a large number of impartial witnesses, chosen specifically because of their 'advanced years' and their consequent ability to draw on personal experience and historical knowledge of water management and previous flooding in the area. While witnesses 'two, six, eight and twelve' suggested that the cause of the floods were the intakes and dam which had been constructed, the other witnesses argue that the small river channel, the poor nature of the river banks and flood walls and the river's dominant situation with respect to the city were more significant factors. Indeed, if anything, the water systems were argued to be quite useful in that they 'abstracted water from the general flow of the river' and it is suggested that 'before they were there, the river had more power' and hence posed more of a flood threat.[44] Evidence drawn from personal experience and social memory of severe and recurrent flooding is thus employed to suggest flooding was, and had been for some time, a persistent problem in this area, but that water management was alleviating, not causing, the problems. Thus, while communities affected by flooding argued that the manipulation of water from the river and the damming of the channel by local hacendados had increased the flood hazard, the hacendados themselves, together with a series of supposedly impartial witnesses, suggest that the water management had, if anything, served to lessen the problems of flooding in the region.

To resolve the differences, the local council employed Jose Mariano Orinuela, described as an 'expert' in the field of flood risk, to explore the problem and review the evidence. Following a detailed survey of the region and the water management structures, he concluded:

> There is no doubt that works ... to extract or deter the waters of the river should be undertaken only with special precautions especially when there are populations located nearby. Drainage or irrigation channels crack the ground immoderately where they pass, bringing much material which is discharged into the channels and flows down ... inundating the sown plots of land, houses and roads: the dams and reservoirs collect deposits of sand and mud that they build up on the bed of the river reducing its capacity and in times of high water, causing the water to pour out across the margins of the river banks causing damage to the neighbouring lands.[45]

Expert opinion thus effectively supported the case forwarded by the indigenous community, though sedimentation was highlighted as one of the key factors exacerbating flood risk. To some extent, the predominantly anthropogenic explanations of flooding may reflect a general awareness of the power relationships between different social sectors of the population, the level of differential social vulnerability to drought and flooding, and the potential gains that could result by proving a party had acted unlawfully. By challenging neighbouring landowners' use or misuse of water sources, or charging them with damages due to dam neglect and flooding, fines could be levied and water rights modified.

Problems were not of course restricted to the Bajío. Water management structures created problems for householders in San Felipe el Real de Chihuahua. In 1729, for example, Don Juan Sanchez de Comancho, Sargento de las Reales and Corregidor of the town, undertook a survey of the water channels constructed by and belonging to Juan Perez, resident of the town. The document indicates that these channels were causing problems for local resident Don Felix de Ochoa, whose house, along with those of 'other residents immediately next to the said channel', had suffered considerable damage. The problems lay, it seems, in the 'continual increases' which the river suffered during the wet season, the impacts of which appear to have been aggravated by the building of the channels. The survey resulted in Perez being asked to 'relocate the said channel where it was not hazardous or to build a fence or tanked wall for the said channel'.[46]

Flooding was apparently a more regular problem for Dona Joana de Sanoa, widow of Don Francisco Gonzales de Celis, and her son, Don Antonio de Celis, residents of San Felipe el Real de Chihuahua in the 1730s. It appears Dona Joana's son was being fined 25 pesos and was being threatened with incarceration for eight days for blocking the water channels and ditches belonging to Don Juan del Rincon. In her son's defence, Dona Joana appealed to the moral conscience of the judge, calling upon the need for the 'justice and defence of poor widows', and claimed that the drainage channels and water infrastructure had repeatedly led to flooding which had damaged her property. Supporting her son's actions, she noted that the water served 'no common benefit' and was destined only for a garden plot belonging to Rincon.[47] Only two years later, there were similar problems recorded in San Felipe el Real de Chihuahua once more, though this time in relation to the construction of a new Jesuit college. According to one resident and mine worker, Don Jose de Vegrar, the construction works had modified the normal flow of water, causing flooding in the houses nearby, some of which were noted to be very old indeed. 'The natural and ancient course of the waters had not caused any damage to any citizens.' The new development, however, had resulted in 'gross damages' to Vegrar's house and to those of two other residents. Indeed, it had, he argued, become impossible to use the existing entrances and exits of his house.[48]

Repeated flood events, particularly those in more rural contexts, may have also presented opportunities. In Oaxaca for example, it was not unusual for litigious indigenous communities to seize upon flood events to gain territorial advantage. The sequence of maps showing part of the River Atoyac and environs in Oaxaca (figures 5.1–5.3) illustrates one such instance. The maps were produced to show the impacts of a series of flood events between 1775 and 1804 on the River Atoyac in Oaxaca and were used as supporting evidence in a land dispute between the community of Xoxocotlán and the estate owner of the hacienda of San Miguel. The two parties used the river as a territorial boundary between their respective lands. That this part of the river was prone to flooding is evidenced by the extensive flood defences constructed on the south bank of the river, and a substantial floodwall, 4 varas (just under 4 meters) high, erected by the community of Xoxocotlán. Following a flood event in 1775, however, there was an avulsion event,[49] or a shift of the river channel, as figure 5.1 illustrates. Annotation on the map notes that 'the channel that from time immemorial has been for the Atoyac River up to the year past of 1775 ... was moved by a tempestuous flood'. The channel in the centre of the map is described as the 'new channel of the river and where the river has flowed since 1775 and does so presently'. The flood event thus resulted in substantial landscape changes.

The community seized upon this event, together with continued avulsion of the river channel between 1775 and 1785 (figure 5.2), to forward a case for a redefinition of the boundaries between their lands and those of the neighbouring hacienda. According to the hacendado, the community, 'against all reason', attempted to 'meddle with the boundaries'. Interestingly, as figure 5.3 illustrates, the migration of this part of the Atoyac, and hence also the degree of contention between the two

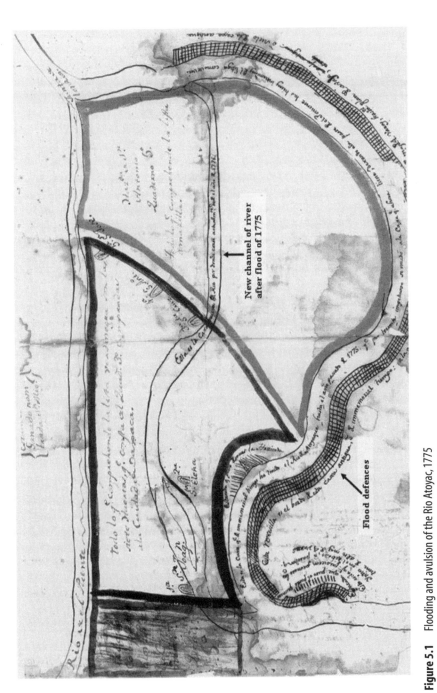

Figure 5.1 Flooding and avulsion of the Río Atoyac, 1775

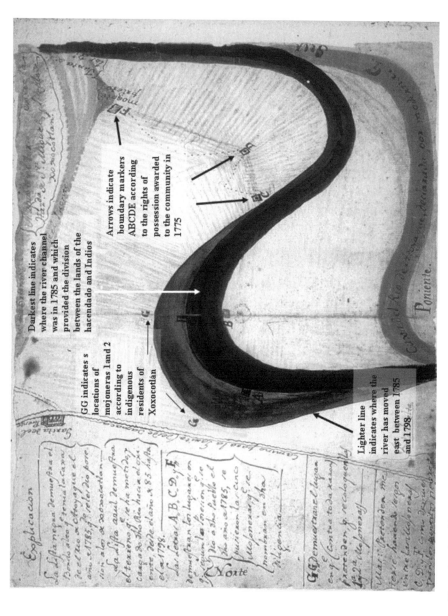

Figure 5.2 Disputed territorial boundaries in 1785 following avulsion event

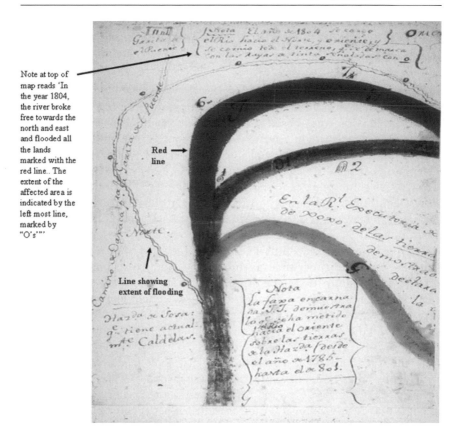

Figure 5.3   Further movement of the river channel and flooding in 1804

landowners, would continue into the early years of the 1800s, when, as will be considered in the following chapter, unusually wet conditions led to another period of flooding. A note at the top of the map suggests that 'in the year 1804, the river broke free towards the north and east', not only flooding an extensive part of the property belonging to the hacendado, but effectively prolonging the dispute between the neighbouring landowners.[50]

Inasmuch as droughts and floods might have been instrumental or were implicated in driving some of the more localized legal conflicts over land and water, there are examples of more widespread and violent unrest during the colonial period which may have links to climate variability and weather events. The following section explores the possible relationship between prolonged drought and social unrest in the north of Mexico.

Regional Resistance: Drought, Disease and Rebellion in Northern Mexico

In Chihuahua, the degree of disquiet over climate variability, drought and its implications, or more specifically its consequences for communities whose livelihoods were based predominantly on hunting and gathering and/or subsistence, might have extended beyond legal conflict and localized social unrest. As mentioned in chapter 2, this region was bellicose. In-fighting between the very many different cultural groups in the north of Mexico was commonplace in the pre-Hispanic period and there was more or less continuous warfare with the nomadic groups in the region dating back to the first stages of European contact. As Deeds (2003: 1–2) suggests, indigenous concerns there, more than any other region, centred on 'the struggle to preserve cultural autonomy, the need to obtain sustenance from the land and other natural resources ... and recurrent battles with disease and malnutrition. The strategies for survival were multiple and included rebellion.' This potential for violent rebellion posed the most serious threat to Spain's hold on the northern frontier of New Spain and underscored the need for almost constant Spanish military presence across the region.

While the intensity of actual warfare that took place on the frontier varied, hostilities and fighting were fairly constant (Griffen, 1979). The severity of the raids on occasion necessitated the temporary abandonment of some of the early colonial settlements. In 1586, the colonists were forced to leave Santa Barbara due to attacks on the town, its colonists and the conchería labour force.[51] Attacks worsened in 1594, leading to a request for financial assistance from the Crown to support the defence of the region (Cramaussel, 1990a: 46). Nonetheless, raids continued throughout the seventeenth century, with particularly violent attacks taking place in 1601, 1616–20, 1644–5, 1648–52, 1666, 1684, 1690 and 1697.

The perpetrators of hostilities would change throughout the course of this period. While, for example, the Acaxees were attributed with the unrest of 1601, it was the Tarahumara and Tepehuanes that caused most problems for the Spanish throughout much of the 1600s. The Tepehuan revolts of 1616–20, for example, in which over 200 Spaniards were killed along with ten missionaries, and an unknown number of slaves and servants, and which resulted in an estimated 400,000–1,000,000 pesos of damage, have been well documented (Gradie, 2000). The violence came at a time when this bellicose group had, it was thought, been pacified, but it tested the strength of Spanish institutions in the area as well the Tepehuan capacity for military and cultural resistance (Gradie, 2000: 8). Tarahumara insurrections throughout the 1600s would also prove to be an obstacle to Spanish colonizing efforts across the region, and particularly in the area to the west and north of the Parral district and to the eastern desert area. Franciscan and Jesuit missions, for example, were attacked during the Tarahumara rebellions of 1648–52. The Spanish defence, led by Juan Fernandez Carrion, responded by setting fire to the fields, so forcing the rebellious Indios into the mountains. Nevertheless, further attacks would follow throughout the 1650s. The area north and east of Parral was also a site of resistance in the period between 1644 and 1645, though this time the offenders in the hostilities were thought to have been Tobosos and Salineros groups. By this stage the Tobosos were thought to have acquired horses, rendering their attacks more violent (Aboites Aguilar, 1994).

In 1645 there was a rebellion in the missions of San Francisco de Conchos. Two Franciscan missionaries were killed, further attacks followed and the mission was razed to the ground. Conchos from the Valley of San Bartolomé united with those of Parral and began to attack haciendas, killing various labourers and stealing livestock. Other rebellions in the Jesuit missions of Salveto and San Lorenzo followed, though there were no killings, the missionaries having already fled in fear. The San Francisco mission would not be repopulated until 1667 (Aboites Aguilar, 1994).

In August 1680 there began one of the most important phases of indigenous rebellion in the entire colonial period. Beginning in Santa Fe New Mexico, different groups of Conchos, Tarahumaras and Tobosos coalesced in violent opposition to the Spanish, their exploitative economic goals and their religious purpose. Indeed, New Mexico was abandoned by the Spanish in this period and in 1680 the newly developed mission at San Francisco de Conchos was once again destroyed, along with that known as Nombre de Dios. In the middle of 1684, the groups in northern Mexico, from the Sierra Madre and the desert country to the east also rose up against Spanish authority. Almost all of the different groups of the region are thought to have been involved: the Conchos, who previously helped in the Spanish defence against the Tarahumara attacks, the Julimes, the Las Juntas and the Sumas. The Tarahumaras, though reportedly 'restless' during this period, are thought to have been the exception (Griffen, 1979: 9). The Conchos, possibly influenced by the unrest documented in the Tarahumara country at the time, also rebelled in the 1690s (Griffen, 1979: 20–21). By the turn of the eighteenth century, uprisings appear to have been less frequent. By this stage it is thought that most of the Conchos had disbanded, been assimilated, employed on haciendas and in the mines or lost to epidemics (Aboites Aguilar, 1994). However, raids, thefts and attacks would continue throughout the 1700s far into the Conchos area, this time attributed to southward moving Apache groups (Griffen, 1979).

A number of plausible explanations for these periods of unrest and rebellion have been forwarded. Most centre on oppression, forced labour, mistreatment, and poor working and living conditions, but also highlight the significance of indigenous 'revolt' as a form of political and cultural resistance. As Deeds (1989), Cramaussel (1990a), Martin (1996) and Coello (1989) have all demonstrated, there was considerable pressure on the indigenous groups and particularly the Tepehuanes to satisfy increasing labour demands as mining activities expanded across the region. Missions became particular targets for attacks given their role as suppliers of labour. An alternative explanation positions the revolts as an expression of 'cultural revivalism' (Gradie, 2000: 4) to reassert religious and cultural life in the face of increasing loss of control to the Spanish, Franciscans and Jesuits in the region. In some situations conflict was associated with missionary civilizing endeavours and the bringing together of groups with longstanding histories of opposition to each other. One document from 1640, for example, charts the conflict that had arisen in Parral, where Conchos had been forced to congregate in a Jesuit mission for Tarahumara Indios. As Captain Juan de Barranza noted, 'the Indios had always been enemies and when they had been introduced to the Tarahumaras, they invaded and seized their maize plots and lands'.[52]

Yet there is an alternative explanation for some of the instances of unrest. The periods of indigenous revolt in the north of Mexico in the middle of the sixteenth century (between 1540 and 1580), for example, have been linked to a severe and prolonged drought (Cleaveland et al., 2003: 370), which would have resulted in extreme conditions for all sectors of society, but especially those whose existence was based largely on hunting and gathering. Other periods of unrest may similarly be associated with the vicissitudes of climate and its implications, specifically famine and epidemic disease, in this more marginal part of Mexico. Drought and pests, for example, had damaged the corn harvests in 1615, immediately preceding the Tepehuan revolts at that time. Deeds (1989: 435) has highlighted how the Tarahumara staged their first revolt after a smallpox epidemic in 1645, while other revolts in the 1640s, 1650 and 1652 came after several years of drought and successive new epidemics (Aboites Aguilar, 1994; Deeds, 2003). Epidemics in 1647 and 1662 are thought to have contributed to unrest during these periods and most certainly led to a high level of population mobility. Increased raiding by nomadic groups who had been settled in various mission stations, particularly in the sierra, are thought to have in part been a response to drought, famine and disease during the 1670s, 1680s and 1690s (Deeds, 2003: 86).

In the 1690s, a suite of factors may have combined to precipitate crisis and unrest across the region. Drought in the first three years of the decade was followed by shortages of wheat and maize. These events were followed by epidemics, including an outbreak of smallpox in 1692 and 1693, succeeded by measles, the combined impacts of which led to the loss of one third of the indigenous population of Nueva Vizcaya (Deeds, 2003: 96) and may have rendered existence in this marginal environment even more difficult and so contributed to the repeated Tarahumara revolts of this period (Deeds, 1989: 439). This was, however, also a time of increased Spanish activity in the area, consequent upon the opening of the Parral mines. There was an influx of people and the number of Spanish landholdings increased across the region. The increased draws upon the land, food and water and other natural resources combined with rising demands for repartimiento labour placed pressure on the Indio groups of the area and might have contributed to the revolts. According to Griffen (1979), there are accounts, however, which refer to a great drought in the Casas Grandes district at this time, so drought, and its repercussions, cannot be ruled out as plausible explanations for this period of unrest. Either way, by 1698, the scarcity of food produced across the region, the destruction of the crops and the general debility as a result of drought, rebellion, epidemic disease and economic dislocation had reduced the level of indigenous resistance. By 1699, Juan Fernandez de Retana, the leader of the Spanish defence, was able to restore peace across the region (Aboites Aguilar, 1994).

Peace, however, was shortlived and indigenous attacks resumed in the 1700s and escalated in the second half of the century. While much of this unrest is thought to have stemmed from population expansion, exploitation of indigenous labour and pressure on limited land, mineral and water resources, analysis of the archival documentation in Chihuahua indicates that later periods of prolonged and severe drought may have played a critical role in triggering repeated attacks, raids and also livestock thefts. Figure 5.4 presents a summary of archival references to drought, subsistence crisis and famine, epidemic disease and unrest in the eighteenth century. While it is clear that numerous factors contributed to periods of unrest and it would be foolhardy to draw direct relationships between drought, dearth and epidemic disease (Macleod, 1973), there do appear to be periods when circumstances coincided to render some sectors of society less able to cope and more likely to resort to violence.

Harvests failed between 1724 and 1727, resulting in maize shortages and scarcities of wheat flour.[53] There were also concerns over the availability of meat in the Villa de San Felipe el Real de Chihuahua at this time,[54] and although, as illustrated in the previous chapter, some sectors of society resorted to begging for food, others were driven to theft. There are in fact many reports of raids on livestock ranches in the region at this time.[55] In 1738, once again in a context of 'extreme drought' which had ruined the crops in that year, there were 'Indio depredations as a result of the hunger that they suffered'.[56]

In the latter half of the 1700s, nomadic groups raided far into the region and violent attacks became more numerous (Griffen, 1979: 24). From the

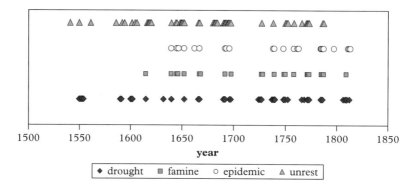

Figure 5.4   Droughts, famine, epidemic disease and unrest in Chihuahua, 1550–1820

late 1740s and early 1750s on, for example, the parish registers of Chihuahua record an increase in the number of burials of people killed in indigenous attacks (Martin, 1996: 27). By this stage, however, it was the Apaches who were held responsible. Drought in the years between 1749 and 1751 in Chihuahua created extreme hardship and because of the 'great scarcity' at this time, few people had been able to escape 'the pernicious calamities that occasion … hunger'.[57] 'The sterility' recorded in 1749 had resulted in there being only limited pasture available for livestock grazing,[58] while further drought in the region in 1750[59] contributed to a lack of seeds, 'especially maize',[60] and resulted in the death of livestock.[61] The 'urgent need' and 'scarcity of grazing livestock' seem to have reached crisis point by 1751, as there were apparently insufficient numbers of livestock from which to breed.[62] Moreover, the problems continued into 1752, when there were reports of 'cattle dying in the field'. Indeed, it was said that 'the death of the livestock had come following the rigorous drought … the same drought has occasioned in the livestock an illness'.[63] A document from February 1750 highlights how the 'rigorous' drought conditions that year may have incited 'the contention of the enemies',[64] while attacks in 1752 were directly associated with the problems which were experienced across the whole area as the drought continued. 'The decay that each day grows in this town is the cause of the hostilities that the barbarous Indian enemies occasion with the deaths of people, the theft of horses and mules.' Livestock thefts and deaths were in turn affecting the ability to transport wood, charcoal, metals and materials[65] and led to a disruption in local mining activities.

Hostilities continued into the middle of the 1750s,[66] and may again be linked to drought related harvest failures, food scarcity and its implications. By January 1758, for example, the 'lack of maize for daily consumption' in

San Felipe el Real de Chihauhua,[67] and the limited reserves in the alhóndiga, might have contributed to the 'general pestilence' between 1758 and 1759, which resulted in a third of the population being buried.[68] The epidemic is described as comprising a 'contagious fever', but there appears to have been some debate as to whether the illness represented an outbreak of tabardillo (typhus) or another disease. What is clear is that 'it affected entire families, sparing neither small nor large, young or old people'. Population figures indicate a combined estimate of between 6,000 and 7,000 people in San Felipe and Santa Eulalia in 1759, though it is interesting to note that records from this period suggest that epidemic disease during the previous three years had already reduced the population significantly (Martin, 1996: 26). Though the period of disease might not be related directly to the drought conditions of the 1750s per se, food and water scarcities made for a much weakened and susceptible population and it may be no coincidence that this period witnessed 'repeated invasions of enemy Indios', with some groups 'robbing mules and horses, killing them and committing other atrocities'. In fact, the theft and killing of livestock was cited as one of the key reasons for there being insufficient grain being transported to the alhóndiga at this time,[69] and could account for some of the documented scarcities noted earlier. Haciendas and ranches were abandoned (Martin, 1996: 25) and a small group of merchants began to collect contributions for a campaign led by Governor Juan Francisco de la Puerta y Barerra against the raiding Apaches.[70]

Notwithstanding local attempts to defend their properties, more attacks followed in other periods of harvest shortfalls. After a 'sterile year' in 1763, problems of food scarcity descended on the region once again.[71] Drought affected the poor indigenous population most severely and records indicate that because of the 'very grave necessity' and hunger at this time, some of the poorest sectors of society offered to work on the haciendas and in silver mines in return for food and shelter.[72] Only a year later in 1764, however, food scarcities had contributed to 'the death of diverse people of all classes, especially the poor and devalued people who lack food',[73] while a document from 1765 indicates that the 'lack of wheat flour, maize, money and buyers of the little seed that is available' had been exacerbated by harvest losses due to 'the hail of the previous year'.[74] In a letter written from Parral in 1766, a direct link is drawn between this period of apparently extreme hardship and the resumption of indigenous attacks. The author relates the 'notable damage in the haciendas because of the rigorous drought' of that year and highlights how 'the enemies and thieves rescind on their agreements ... and have begun ... to satisfy and contend the creditors with the same goods that they have stolen'. The 'goods' referred to included houses, crops and livestock.[75] Attacks and thefts were thus resumed at a time when unusual weather events had devastated harvests and caused acute social and economic distress. In fact, a number of haciendas, such as the Hacienda of Palo Blanco, were completely abandoned in this year, 'for fear of the Indios' (Lafora, 1958).

Severe and continuous drought reappeared in the first three years of the 1770s.[76] As noted in the previous chapter, poor wheat harvests at this time had led to a shortage of flour and high grain prices, which in turn conditioned a scarcity of bread and there was concern over the implications this scarcity had for the servicing of the mines and haciendas.[77] In 1772, in the middle of this period of hardship, 'flying squads' of Spanish militia were being enlisted to provide some form of defence against the repeated indigenous attacks.[78] Documents indicate, for example, that there was a need to guard grazing livestock at this time in an area known as Laguna de Castillo on the outskirts of the city of Chihuahua. An argument was made that 'there was no other grazing pasture because of the severe drought' that was being experienced.[79] All such defensive activities, however, were impeded by the drought itself. One letter from September 1772, sent to Viceroy Don Anotnio Bucareli y Ursua from Colonel Hugo O'Conor,[80] indicated that 'ultimately, because of a drought so terrible and general ... until now not even a small rain shower has taken place and because of this the fields are all kept unfortunately dried out and all the watering holes have dried up and this is totally adversely affecting the conceived plans.' O'Conor also speaks of a number of ranches that have witnessed livestock thefts and describes how mule trains carrying goods from Michoacán, in west central Mexico, to the north of the country had been attacked, the mules stolen and their cargoes left in the road.[81] Other letters at this time mention further problems that the defending troops were facing. The 'terrible drought' recorded in a document dated 7 July 1772, for instance, had also hampered the herding of livestock such that large numbers of animals had gone missing.[82] Moreover, the scarcity of water was impeding 'every action because there were no horses able to provide power',[83] and a year later, the 'terrible state of the livestock' due to the 'drought of the previous year' was still being lamented. Moreover, the horses that the defending troops had received were argued to be 'young and untrained' and 'because of the drought that they had experienced, they did not have enough energy to fight the enemies that were attacking the ranches and villas'.[84] By the following year, the horses were 'in a very unfortunate state, due to the suffering from the terrible drought of the year before and due to walking 200 leagues because of the lack of grazing pasture, because it was winter and any water was two or three days away'.[85]

The desperate years of drought and famine that affected the region in the mid-1780s may have once again rekindled the activities of 'barbarous indians'. Indeed, one document from 25 August 1787 even suggested that the 'cruel hostility' was far worse at this point than at any other time, and given the geographical scale and severity of what would unfold into one of the most devastating periods of agrarian crisis, it is perhaps not surprising that retaliations and attacks were indeed more aggressive.[86] Changes were, however, made to the Spanish policy towards the semi-nomadic and nomadic indigenous groups around the 1780s. The colonial administration began to grant them food and cattle subsidies to encourage their settlement and so lessen the risk of raids (Katz, 1988). Although links between the free peasants of the north and the colonial administration were no doubt strengthened by such policies, attacks on settlements, military garrisons and livestock would continue into the first two decades of the nineteenth century and would contribute to the abandonment of some of the larger haciendas in the region (Curtin et al., 2002). Cattle theft in particular continued to be a problem following independence and is thought to have been among the most prosecuted crimes in the whole state of Chihuahua by the later decades of the nineteenth century (Lopez, 2001).

Not all of the attacks documented in the archives were necessarily associated with discrete drought events. Increasing social tensions and dissent across the country in the later eighteenth and early nineteenth centuries among the poorer, disenfranchised and more marginal groups of society may well account for some of the instances of unrest. Northern Mexico, however, has had a longer history of conflict. The struggle for subsistence by the indigenous populations of the region may explain some of the more brutal battles of the sixteenth century discussed elsewhere (Naylor and Polzer, 1986). Drought, however, might have exacerbated the general difficulties of a nomadic existence on the northern frontier, explaining at least some of the attacks in the later colonial period.

Vulnerability, Riots and Rebellions: Rare Events or 'Tipping Points'?

Meteorological fluctuations both directly and indirectly affected and influenced a variety of activities in colonial Mexico. Pressure on food supply following drought induced harvest failures, for example, may have contributed to out-migration and the dislocation of some rural indigenous communities, especially those occupying more marginal environments or areas of ecological vulnerability. There was increased mobility and specifically more rural to urban migration. This in turn may have also generated varying levels of unrest both in the host and destination communities (Hamnett, 2002: 77). The municipal governments of Mexico were relatively effective in providing basic food supplies during all but the most critical periods of subsistence crisis. The provision of positos and alhóndigas ensured, for the most part, that even the poorest sectors of society had access to basic foodstuffs during periods of food scarcity and helped hedge against unreasonable inflation of grain prices at the hands of capitalist merchants and traders. Such systems though were less effective during prolonged periods of food shortage, and were only partially successful in averting problems of speculation and profiteering. As Cope (1994: 130) has suggested, 'even abundant supplies were of little use to the urban poor if they could not afford to purchase them', and as has already been illustrated, speculation and supply problems were not uncommon.

On some occasions such problems combined and may have contributed to violent urban rioting, often coincident in multiple foci. The time between 1690 and 1692 was, for example, particularly trying (Cope, 1994: 125). As shall be demonstrated in the next chapter, a series of calamities combined to ruin both autumn wheat and maize harvests.[87] Prices rose dramatically and famine and disease in epidemic proportions then gripped many parts of the country, causing massive life loss, especially among the indigenous populations (Orozco y Berra, 1938; Berthe, 1970; Van Young, 1988: 143; Cope, 1994). Urban riots took place in Mexico City in 1692, involving several thousand people. There was looting and extensive property damage and a number of deaths were recorded (Cope, 1994). Rioting was also reported in the Bajío at this time (Marmolejo, 1967: 181), where drought, food scarcity and epidemic disease were followed by other calamities. The flood in Celaya which opened this study, for example, exacerbated these more widespread prevalent problems and represented something of a 'tipping point', contributing to or even triggering the emergence of popular uprisings (Marmolejo, 1967).

The events of the first couple of years of the 1690s may have represented in Scott's terminology a 'shock … of substantial scope', such that a large body of the populace, with a collective reason to act, did so. It was simply too difficult to adapt routinely or incrementally under the combined circumstances of food scarcity, famine and then flood. As Scott (1976: 193) observed, however, 'to speak of rebellion … is to forget both how rare these moments are and how historically exceptional it is for them to lead to successful revolution'. Riots, rebellions and instances of unrest in both the city and the country were, in the main, unusual. Most riots, demonstrations of resistance, attacks and incursions in both urban and rural contexts remained relatively localized and short lived. Urban protests were often dealt with through 'a calculated mix of repression and negotiation' (Serulnikov, 1996: 192) and as a result, for much of the time, the state managed to retain control, while ensuring a sense of justice. The majority of protestors normally remained 'fairly convinced that the Crown … would redress their wrongs' (Katz, 1988, cited in Serulnikov, 1996: 193), while, largely as a result of the demands placed on individuals by an agricultural way of life, periods of rural unrest also tended to be short lived. All too soon rural protestors were required to attend to the fields.

All the examples of unrest highlighted in this chapter, they should be seen as part of a rich tradition of political and legal engagement in Mexico. Centuries of elite centred histories portrayed rioters and rebels as irrational or at best as faceless victims (Arrom, 1988: 2). Following a line of argument forwarded by Hobsbawm and Rudé (1968), however, Mexican rioters, just like European ones, were far from irrational. Motivated by a shared set of beliefs and grievances, rioting or protesting was a means of political expression, presented a very real (if, in most cases, only temporary) threat to government control of some major cities and was underscored by racial tension and political dissent.

The elements for galvanizing common action in different parts of the country, that is to say, the causes of widespread or nationwide insurrection, remained absent, at least until the late eighteenth century and the early years of the 1800s. Up until that point, a variety of individual, collective, institutional, long- and short-term, adaptive and survival strategies, as illustrated in the previous chapter, managed to stave off the opportunities for countrywide rebellion. Throughout the eighteenth century, a series of socioeconomic, demographic and structural changes took place, modifying the relative vulnerability of different sectors of society to climatic variability and unusual weather events and indeed exacerbating the level of vulnerability for some. Yet there was also an increase in the level of environmental awareness and cumulative knowledge of climate variability and its potential implications, informed by social memory and previous experience of different types of weather related disaster. The way in which people manifested and articulated grievances about these differential social vulnerabilities also changed. By the later part of the century, a suite of social, economic, political and environmental factors combined to produce an untenable situation for the majority of the population. This was coupled with a growing consciousness of inequality, exploitation and repression, together with opportunities for acting upon this in a non-peaceful way. For these reasons, this period witnessed the galvanizing of common action against the governing regime, mobilized by an intellectual elite (Hamnett, 2002). The following chapter explores the part played by environmental and climatic developments in the anatomy of this period of popular unrest.

Chapter Six

Illusory Prosperity: Economic Growth and Subsistence Crisis in the Disastrous Eighteenth Century

Introduction

As Van Young (1993) has suggested, the eighteenth century in Mexico was at once a time of prosperity and depression. This was undoubtedly a period of impressive demographic growth and economic dynamism, fostered by particular expansion in both the mineral and agricultural sectors. As a result, this period has been highlighted as something of a golden age of economic triumph. Towards the later part of the century, however, mining production and output began to slow down and the demographic vitality that had so characterized the first part of the century showed signs of waning.

A series of dramatic subsistence crises linked to meteorological anomalies may have contributed to – or at least compounded – the difficulties. Indeed, it has been suggested some periods of crisis may have contributed to the unrest that characterized much of the late eighteenth and early nineteenth centuries (Florescano, 1986). Periods of harvest shortfall, food scarcity and famine could have implications for all sectors of the economy. Invariably, if drought had been the cause of harvest shortfalls and problems in the agricultural sector, pasture was another victim, disrupting the livestock economy. There could be other more indirect effects, too. Mules were critical to providing power to run refineries and yet were commonly fed on maize. Shortages of this most fundamental crop could, therefore, also have implications for the functioning of the mineral sector. In fact, as Hamnett (2002) suggests, a full blown subsistence crisis could lead to a suspension of operations in the mines, leading to large-scale unemployment and thus also migration. Migration, in turn, might lead to increased vagrancy and higher levels of urban crime, infractions and social disharmony.

As has been suggested elsewhere (Florescano, 1986; García-Acosta, 1993, 1997; Endfield et al., 2004a, 2004b), there were numerous subsistence crises linked to unusual and extreme weather events throughout Mexican history and prehistory, with complex and often interlinked implications. A number of authors, however, have highlighted a series of particularly devastating, widespread agrarian crises in the eighteenth century, most of which appear to be associated with anomalous weather. The crises of 1710, 1724–5 and 1741–2, for example, were preceded by drought and harvest shortfalls, while that of 1749–50 was associated with disastrous frosts which ruined the sown crops in western and northern parts of the country. Between 1771 and 1772 (and possibly including 1773 in the north of the country, as already shown in chapter 5), there was prolonged drought, which led to harvest shortfalls and high grain prices. Problems in 1781–2, meanwhile, were preceded by three years of drought and epidemic disease (Cooper, 1965), while the great crisis of 1785–6 was, as suggested, the result of a combination of delayed rains in the first few critical months of the newly sown crops, succeeded by two exceptionally severe frosts which destroyed the harvests in the centre, west and north of New Spain (Florescano, 1972: 56). Swan (1981) highlights a general rise in the price of grain round about the turn of the nineteenth century, which she links to abnormal weather patterns and which affected crop production across the agrarian heartland of central Mexico. In fact, as will be shown, a variety of climatic perturbations affected the harvests between 1801 and 1804, while the crisis of 1810–11 was the result of a terrible drought which lasted for the whole of 1809 and 1810 and affected all cereal producing regions of the country (Hamnett, 2002). Moreover, the 1700s began after one of the most calamitous decades in Mexican colonial history, the implications of which must surely have influenced social and economic well-being in the first years of the new century.

The crises of the eighteenth century were of course not unique in any of these respects. But periods of anomalous weather in the 1700s, and particularly those that occurred in the later part of the century, exposed the extreme social and economic inequality that characterized much of the country by this stage, and in so doing highlighted the spatially and culturally differential vulnerability of Mexican society. This chapter will explore the anatomy of a number of these periods of widespread crisis and will examine how they unfolded in the three case study regions of the country. It will be shown that the periods of crisis may have been more complex and lasted much longer than the meteorological phenomena that contributed to them, and that there were regionally distinctive impacts and implications. In addition, the dramatic impacts of a number of little studied and possibly hitherto unidentified periods of anomalous weather and extreme events will be considered. It is first important, however, to provide a broad outline of the preconditions for crisis, that is to say, the social, economic, demographic and political contexts within which these events took place.

Decline and Depression in Seventeenth-Century Mexico

Much has been written about the 'general crisis of the seventeenth century', a time of deteriorating climate, harvest problems, economic depression, political instability, and upheaval and rebellion across Europe (Parker and Smith, 1997). Spain was no exception to this general trend. From the late sixteenth century, the Castillian lands had begun to slide into an economic depression. Plague, epidemics, drought induced harvest failures and other agrarian problems, not least the continued disputes over pastoral grazing rights and the loss of peasant independence to urban capitalists, together with war, combined to affect primary and secondary sector production across the country. This established a trend of economic decline that would persist for another century and a half, affecting Spain's trading position in the world (Israel, 1975; Kamen, 1991; Thompson and Yun Casalilla, 1994).

Whether economic decline prevailed elsewhere in the world has been a subject of debate (Parker and Smith, 1997). It has long been suggested that for the fifty years between 1620 and 1670, however, Mexico might also have been in the grips of an economic depression (Chaunu and Chaunu, 1955; Stein and Stein, 1970). Several hypotheses have been forwarded to explain the nature and dynamics of this phenomenon. Severe labour shortages in the wake of indigenous depopulation and a decline in silver production are thought, by some authors, to have led to a period of depression across the country (Borah, 1951; Chaunu and Chaunu, 1955; Macleod, 1973). Silver became the most important commodity produced in New Spain and was pivotal to the success of the new economy. Its apparent decline in the seventeenth century is thus assumed to have contributed to overall economic depression in New Spain. Certainly, after an initial flurry of activity in the early colonial period, silver mining activities began to decelerate in the 1620s. Labour shortages in the mines, consequent upon indigenous depopulation, served to increase production costs, as did the exploitation of the deeper mineral veins when the near-surface resources had been exhausted. Other arguments have agreed that there was a general economic crisis, but that this was less a function of labour shortages and more a reflection of increased fiscal demands on the evolving colonial political economy, together with attempts to eliminate corruption and hence prolonged political tensions (Israel, 1974).

Apparent economic decline in the colony may have also been a possible function of the coincident economic crisis in Europe. There is no doubt that the weakening of the relationship between Spain and Mexico, and specifically the duration and severity of the depression in Europe, had important consequences for the colonial political economy (Knight, 2002). By the 1630s – when the impacts of the economic recession began to be felt – the Crown imposed more restrictive legislation on the distribution of mercury – an essential commodity in the production of silver. Previously, mercury had been distributed on a credit basis, leading to impressive production figures while at the same time hiding the indebtedness of the miners. The new tighter controls on distribution meant that many miners who had accumulated debts were ruined by demands for repayment. Consequently, by the middle of the century the industry was controlled by financiers based in Mexico City (Brading, 1971), while some of the more marginal producers were forced out of business and had to abandon their mining operations altogether (Boyer, 1977). In addition, there were other indicators of economic depression. According to tax and tribute records, output of wheat from the Bajío, for example, declined between 1669 and 1696. To some extent, however, this decline is thought to have been indirectly linked to declining mineral production. Producers may have wanted to avoid the risk of overproduction and so planted less (Boissonas, 1990: 132).

Other authors have questioned whether a period of economic decline prevailed at all in New Spain. Echoing ideas first propounded by Gunder Frank in the late 1960s (Gunder Frank, 1969), it has been suggested that this was a period during which the colony, out of necessity because of the economic depression in Europe, became far more self-sufficient. Although the few indices that are available to assess activities in the commercial and agricultural sectors do suggest a possible period of stagnation, mineral production might not have declined per se. The primary cause of any obvious decline in silver revenues might well be explained by the creation of new provincial treasury districts.[1] Silver output actually declined less than silver exports, or as Knight (2002: 177) suggests, 'the colony retained a larger share of its own most valuable commodity'. Moreover, although international trade might well have declined, the colonial economy itself witnessed a release from mercantile control and began to focus on a growing domestic demand and a growing intercolonial trade (Fisher, 1997; Knight, 2002). Rather than a period of crisis, therefore, this may have been a time of economic reordering (Florescano and Sanchez Gil, 2000). Furthermore, Boyer has highlighted how Mexico City assumed a more independent metropolitan role as shipping between Seville and the colonies became less frequent. The capital came to represent the administrative hub of trade with the New World as well as with Spanish interests in the Far East (Boyer, 1977). While it is clear that the second half of the century was not a period of spectacular growth, this was a time when new settlements and agricultural land investments were consolidated. This was a period of population recovery among the indigenous populations and remarkable population growth among the white populations (from 63,000 in 1570 to 125,000 by 1646). There was an emerging social diversity, while a vibrant import and export trade points to a definite vitality which reflected a transition to capitalism, economic diversification and the emergence of a new social order (Florescano and Sanchez Gil, 2000: 476). Perhaps, then, the seventeenth century witnessed a crisis of transition rather than one of depression or stagnation.

Economic Boom and Bust: Absolutism and Globalization in Late Colonial Mexico

If the seventeenth century was a time of transition, however, the eighteenth century was one of consolidation and sustained and spectacular growth (Florescano and Sanchez Gil, 2000). There was undoubtedly a dramatic expansion in population. Between 1740 and 1810, Mexico's population is thought to have doubled from 3.3 million to 6.1 million (Brading, 1978), and in 1800 the population was no less than 2.7 times greater than a hundred years before. In the 20-year period between 1790 and 1810, the population of Mexico increased from 4,483,564 to 6,122,354, most of the growth being in the indigenous, Mestizo and Casta populations (Beltrán, 1946). There was also very impressive economic growth over this period. To some extent this was stimulated by political and administrative developments in Spain. Upon the death of King Charles II in November 1700, the entire Spanish inheritance fell to a member of the French Bourbon Dynasty, Philip V. Having a Bourbon king on both the French and Spanish thrones disturbed the balance of power in Europe and when Charles died, a Grand Alliance of European nations united against Philip in the Wars of Succession (1702–13). In the Treaty of Utrecht signed on 11 April 1713 Philip was finally recognized as king of Spain, but on condition that Spain surrendered its possessions in Europe and allowed British trade with the Americas.

Presented with new threats to the Spanish trading monopoly, Philip took measures intended to counter the decline of Spanish power (Hamnett, 1997). He and his Bourbon successor, Charles III (1759–88), and their administrators attempted to overhaul the machinery of empire and revitalize royal control over Spanish American colonies as part of the Bourbon reforms. These constituted a sweeping array of new domestic and colonial policies intended to revive and strengthen the economies of Spain and her possessions, to reduce local administrative power, improve the army and firearms and encourage industry and trade. As Baskes (2000: 45) suggests, 'no sector of colonial society remained untouched by the reformist Bourbon state'. The reforms entailed the curtailment of ecclesiastical power, reapportionment of colonial territory, the restructuring of colonial military forces and the liberation of trade from the many powerful colonial groups whom, it was believed, constrained economic growth. Indeed, after around 1765, commercio libre, or free trade, was actively encouraged (Lynch, 1992: 111).

The Real Ordenanza de Intendentes attempted to increase the state's control over local affairs by overhauling the colonial bureaucracy (Baskes, 2000: 45). The Real Ordenanza was forwarded by Jose de Galvez, Visitor General to Mexico and Minister of the Indies. As Baskes (2000) suggests, Galvez was highly critical of the enrichment of the Alcades Mayores at the expense of the 'miserable' Indios and the royal tribute.[2] Galvez was also conscious of the growing number of lawsuits that were being brought against Alcades Mayores that local Audiencias had to address. His proposal involved a complete overhaul of the administrative system and the introduction of new political districts or Intendencias. The colony was divided into twelve such districts, to each of which a new Intendente Corregidor was appointed. These individuals were intended to assist the Viceroy in the management and administration of the colony, were answerable to the Viceroy and were intended to be important local level representatives of the Crown (Baskes, 2000). Alcades Mayores were replaced by Subdelegados who were subordinate to the Intendentes and, unlike the former, were salaried, professional bureaucrats.[3] The new system was designed to centralize power, reduce the autonomy of the local district magistrates and so increase central government's control over local affairs (Lynch, 1992: 78).[4]

There were some marked commercial benefits associated with this general period of reform. From about 1720, revenues flowing into the Mexican exchequer increased dramatically, far surpassing, as TePaske and Klein (1981) have argued, sums coming into the Treasury during any of the peak periods of the 1600s. The colonial political economy became even more integrated into the global economic system as Spain followed an economic policy geared towards mercantilism. Silver production increased by a factor of 4.5 between 1700 and 1800 and, as New Spain's most important export commodity, this helped stimulate general economic growth and prosperity (Ponzio, 2005). Spanish migration to the New World accelerated in response to the increase in economic opportunities, together with the need for more Spanish administrators in the colony (Lynch, 1973). Total agricultural production almost tripled during the final decades of the colony and by 1810 had even surpassed mineral output. On appearance, therefore, Mexico seemed to be benefiting from a period of eighteenth-century 'globalization'.[5]

However, it is thought that these signals of growth and export oriented free trade in late colonial New Spain might have masked a number of deeper and longer-term structural problems that would manifest themselves by the late eighteenth and early nineteenth centuries (Ponzio, 2005). Though export oriented areas thrived with free trade policies, Spanish American entrepreneurs were excluded from transatlantic trade and shipping, which the Spaniards controlled. Popular sectors throughout Spanish America viewed new fiscal exaction as endangering their living standards. Sales taxes and the enforcement of state monopolies affected many ordinary traders and consumers. Tribute and forced sales of goods angered Indio and Mestizo peasants and created additional problems. Increased tribute and also consumption taxes forced Indios to sell their labour power for cash, yet demographic expansion meant there was a glut in labour supply. At the same time, and as already highlighted, there was a general increase in the amount of land subsumed under great estates and large landholdings. This placed pressure on indigenous lands and Crown owned commons (Brading, 1978; Van Young, 1981), greatly reducing the capacity of the indigenous subsistence sector and in turn forcing more Indios to seek work as poorly paid labourers (Lynch, 1992) or migrate to the urban centres. There were drives for increased production in key sectors of the economy, particularly in the mining sector and in commerce. Yet there was a lack of capital to invest in infrastructure and new technology to increase mineral production. As a result, this drive for increased output imposed greater pressure and labour demands on workers, but also increased the potential for the main benefits of economic expansion to flow to mine owners and merchants. Agriculturalists and landowners were effectively left behind (Lynch, 1992: 112).

Another associated key objective of the reforms was to undermine the power of the Church. The implementation of these changes took many years and had several effects, but the Jesuits were expelled from Mexico in 1767 and the reforms curtailed the ecclesiastical fuero, an institution that placed the clergy outside the control of civil authorities. A royal decree was issued in 1804,[6] seeking to reduce Church land monopolization, effectively ordering the Church, which had by this time become one of the major landowners and creditors for local hacendados, to deposit in the royal treasury capital held in pious funds and capellanías (Ruiz, 1992). An estimated total of between 10,500,000 and 12,750,000 pesos were seized (Hamnett, 1969). In areas where the Church had grown influential, this measure not surprisingly caused resentment, especially in the lower clergy. Colonials, principally hacendados and rancheros, were, in effect, left without any resources from which to borrow funds. Rich hacendados could liquidate their loans whereas poorer landowners and rancheros could not. Indeed, many haciendas were at this time put up for sale and some former ranches were simply deserted.

The Bourbons also felt it necessary to restructure colonial militias as a safeguard against aggression from other European powers as well as internal social unrest. In contrast to the weakening of the Church, the Bourbon reforms reinforced the military fuero, which placed Spaniards and Creoles, and even people of colour who served in the army, outside civilian jurisdiction. The army created an avenue of social recognition for Creoles and non-white people, and allowed them to develop military skills that later proved useful in the struggle for Independence. The Creole aristocracy felt threatened by policies that allowed non-whites to achieve privileges previously reserved for the elite. The participation of popular sectors, most of them of non-white background, in the Independence conflicts provoked 'caste war' fears, since the elites, considering themselves white, framed any assault on their privileges as a racial attack.

All these forces combined to aggravate tensions between different classes and to increase the vulnerability of some sectors of society. Indeed, Lynch (1992) has argued that the absolutism associated with the reforms might have contributed to the large-scale rebellions in Mexico towards the close of the eighteenth and early nineteenth centuries. A series of anomalous weather events and famines around this time, however, may have served to reveal or highlight this striking inequality and the differential social, economic and environmental vulnerabilities of people across the country, so rendering the colony ripe for rebellion.

'A time of calamity': A Synthesis of Climate and Crises in Late Colonial Mexico

As has been shown in the previous chapters and as summarized in table 6.1, there are many direct and indirect references to droughts, floods, frosts and storms in the colonial records,[7] though not all caused dislocation or contributed to widespread crisis across Mexico. For example, a drought in one year might have negligible impacts if, in the following year, a good harvest could be secured. But if buffer stocks were depleted by one bad season, a second or third in succession could be much more devastating. Successive extreme events, or individual events occurring in combination with other natural hazards or weather events, might thus have led to amplified and cumulative impacts (Wigley, 1985). Possessing little in the way of reserve seed stocks and sufficient, fertile and irrigated lands, it was often poorer sectors of society that were the first to suffer during prolonged crises. By the eighteenth century, however, a whole suite of demographic, social, economic and political conditions coalesced to render society at large apparently more vulnerable to climatic variability and periods of extreme and unusual weather. Indeed, comparing the archival references from all three regions (see table 6.1), a number of particularly widespread and possibly nationwide subsistence crises can be identified in the 1690s, the 1750s, the 1780s and on at least three occasions during the first two decades of the nineteenth century.

Table 6.1 Chronology of recorded weather and weather related events and possible implications in colonial Mexico, 1690–1820

Year	Oaxaca		Bajío/Guanajuato		Chihuahua	
	Weather/weather related events	Other events	Weather/weather related events	Other events	Weather/weather related events	Other events
1690	Limited rains and harvest failure		Drought and harvest failure	Epidemic	Drought/harvest failure	Possible eclipse, 23 August; unrest
1691			Drought (interspersed by summer floods, May, June and August)	Epidemic	Drought/harvest failure	
1692			Frosts/Celaya flood event; urgent need for food in Guanajuato	Urban riots		Smallpox epidemic; unrest
1693						Measles epidemic; unrest
1694	Frosts affect sugar crops					
1695			Frosts; drought		Drought	
1696	Limited rains and harvest failure	Earthquake; chahuistle	Drought/harvest failure		Drought/harvest failure	Earthquake
1697	Frosts affect sugar crops (July)	Chahuistle			Scarcity	Unrest

Year					
1698	Chahuistle			Scarcity	Limited food production; comet?
1700			Chahuistle		
1706			Chahuistle		
1709		Drought/harvest failure	Chahuistle	Drought/harvest failure	
1710		Food scarcities		Food scarcities	
1711			Chahuistle		
1714	Chahuistle and property abandonment		Chahuistle		
1718			Chahuistle		
1721	Possible hurricane, Tehuantepec, May (similar to that reported in September 1599)				
1723	Very heavy summer rains and flooding	Salt losses to flooding			
1724				Drought; 'thin' maize harvest	

(Continued)

Table 6.1 (Continued)

Year	Oaxaca		Bajío/Guanajuato		Chihuahua	
	Weather/weather related events	Other events	Weather/weather related events	Other events	Weather/weather related events	Other events
1725					Drought; harvest scarcities and food shortages	
1726	Drought/harvest failure				Drought; food shortages	
1727		Earthquake			Drought; bread rationing and begging	Epidemic
1728	'Lack of rains'					
1732	Harvest problems					
1733	Drought/harvest failure					Epidemic
1735				Chahuistle		
1736			Unusually heavy summer rains and flooding	Matlazahuatl epidemic		
1737			Flooding	Matlazahuatl epidemic		

1738		Matlazahuatl epidemic	'Extreme drought' and food scarcities (Durango and Chihuahua)	'Mysterious cattle illness'; unrest
1739		Matlazahuatl epidemic	Drought	
1740			Drought	
1741			Drought	
1746	Drought/harvest failure/food scarcity/aid requested	Chahuistle		
1747	Food scarcity			
1748	Food scarcity		Drought/harvest failure	
1749	Drought/frost/food scarcity; food riots	Damaging flood	Drought/harvest failure	Limited pasturage
1750	Drought	Damaging flood	Drought/maize scarcities	Death of livestock
1751	Drought/harvest loss/grain scarcity		Drought/food scarcity	
1752			Drought	'Cattle dying in the field'
1753		Damaging flood	Drought	

(Continued)

Table 6.1 (Continued)

Year	Oaxaca		Bajío/Guanajuato		Chihuahua	
	Weather/weather related events	Other events	Weather/weather related events	Other events	Weather/weather related events	Other events
1754			Drought		Drought	
1755		Etla riots	Drought		Drought (rogation ceremony)	Unrest
1758					Drought (January); lack of maize for daily consumption	'Contagion'; shortages of grazing pasture; 'total ruin' among livestock; unrest
1759					Drought	'Contagion'; livestock death and disease
1760			Damaging storm and flood event		Grain and 'seed' scarcity	
1761				Epidemic		
1762				Epidemic	Food shortages and tribute payments difficult to fulfil	Typhus/smallpox epidemic

Year				
1763	Epidemic		Drought; 'sterile year'; food scarcity and need for grains	Typhus/smallpox epidemic
1764	Epidemic		Drought/frost and hail; hunger and starvation; high mortality (rogation ceremony);	Epidemic
1765			Drought; lack of wheat flour and maize	Epidemic
1766		Drought and food scarcities	Drought	Unrest
1768		Measles epidemic (measles in Mexico City); repeated earthquakes	Drought	
1770		Unusually heavy summer rains/damaging flood	Drought; poor harvests	Unrest

(Continued)

Table 6.1 (Continued)

Year	Oaxaca		Bajío/Guanajuato		Chihuahua	
	Weather/weather related events	Other events	Weather/weather related events	Other events	Weather/weather related events	Other events
1771			Drought/harvest failure	Damaging flood; (earthquake in Mexico City)	Drought; poor harvests and scarcity	Unrest
1772				Damaging flood	'Terrible drought' (July); poor harvests and scarcity	Unrest
1773						
1776			Unusually heavy summer rains and flooding			Unrest
1779		Smallpox/typhus epidemic		Smallpox/typhus epidemic		Smallpox/typhus epidemic
1780			Drought in Guanajuato	Epidemic		Epidemic
1781	'Lack of water'; grain shortages					
1782			Drought/frost and harvest failure; grain scarcity in Léon	Flood (Celaya)		

Year						
1783	Snow, hail, 'evil winds', rainstorms floods/possible hurricane	'Sicknesses'	Drought/frost and harvest failure	'Sicknesses'		'Sicknesses'
1784	Drought/harvest failure		Drought/harvest failure		Drought/harvest failure	
1785	Drought/harvest failure	Epidemic	Drought/harvest failure/famine	Epidemic	Drought/harvest failure/famine (rogation ceremony)	Epidemic
1786	Epidemic	Epidemic		Epidemic	Drought/famine	Epidemic
1787	Very strong wet season and flooding	Salt losses to flooding; 'terrible earthquakes' (April)			Famine	Unrest
1788	Possible hurricane			Damaging flood event (Guanajuato)		
1789						
1790		Earthquake		Flood (Celaya)		
1791	Unusually heavy summer rains and flooding		Unusually heavy summer rains and flooding	Flood (Celaya)		

(Continued)

Table 6.1 (Continued)

Year	Oaxaca		Bajío/Guanajuato		Chihuahua	
	Weather/weather related events	Other events	Weather/weather related events	Other events	Weather/weather related events	Other events
1792			Flooding			
1793	Poor harvests (Zimatlán)		Drought			
1794	Limited rains and food scarcities (Teococuilco)					
1795	Food scarcities (Teococuilco)	Smallpox epidemic/Etla riots				
1796		Etla riots	Hunger (Léon)	Pestilence in livestock		
1797		Smallpox epidemic/Etla riots		Smallpox epidemic		Smallpox epidemic
1799			Frosts and snow			
1801	Drought				Drought	
1802	'Excessive increases' in rainfall and floods	Locusts/crop loss/grain scarcity	Drought/harvest loss		Heavy summer rains (June)	
1803			Drought/harvest loss/food scarcity			

1804	Heavy rains north of valleys in January; flooding; grain scarcities	Famine foods sought; epidemic	Unusually heavy rains/damaging flood event	Epidemic	Drought/harvest failure	Epidemic
1806					Drought/harvest failure	
1808					Drought	
1809	Drought/harvest failure		Drought/harvest failure		Drought/harvest failure	Epidemic
1810	Food scarcities		Food scarcities		Drought/harvest failure	
1811					Food scarcity and famine	
1812					Drought and 'inevitable' food scarcities	
1813				Epidemic	Drought/frost / harvest failure (rogation ceremony)	Epidemic
1814						
1815		Rural unrest				
1816			Drought/harvest failure		Drought	Epidemic

References to flooding have only been noted in the 'weather/weather related events' column if they appear in documents in conjunction with notices of unusual rainfall or storm events. Floods recorded under 'other events' are those which are thought to be associated with anthropogenic factors.

'Unremitting agricultural disaster' in the 1690s

Few decades could be as calamitous as the 1690s. A suite of unusual weather events were recorded across the country. Against a backdrop of, at best, economic transition, and at worst, economic decline, these events were to have dramatic implications for social and economic well-being. In fact, repeated drought in 1690–1, 1692 and 1695–6, frosts in 1692 and 1695,[8] and epidemics, crop blights and social unrest recorded across the country all served to make this decade one of the most challenging in Mexican colonial history. This was recognized at the time. One contemporary account suggests that 1691 was 'memorable for the suffering from the great scarcity of provisions in the majority of the country, the cause being the loss of almost all the harvests the year before' (Marmolejo, 1967: 181). Drought had led to failed harvests in the Bajío and may have contributed to some of the disputes over water recorded at this time,[9] while across central Mexico the drought of 1690–1 was interspersed with heavy rains and flooding in June and July of 1691. Indeed, a number of documents chart the problems associated with unusually heavy, or 'great excesses' of rain during the rainy season of 1691, and particularly during the months of May, June and August.[10]

Such sequential extremes may have facilitated the spread of disease vectors (Acuña-Soto et al., 2002) and may have also created conditions favourable to certain crop blights like chahuistle, possibly accounting for its prevalence in the central and southern regions of the country in the 1690s.[11] At the time this phenomenon was thought to be the result of the combined forces of 'extraordinary accidents', including the particularly wet summer months, but also an abundance of flies, an 'almost total eclipse of the sun' on 23 August 1690 which, it was argued, reduced the light available for the maturing crops, and the fact that 'cockerels crowed in the middle of the night'.[12] Whatever the believed cause, a series of calamities served to ruin both autumn wheat and maize harvests between 1690 and 1691. There were no surplus supplies of grains to mill and the blight thus indirectly reduced the amount of bread that could be made available the following year. The result was food scarcity and hunger across central Mexico. Prices rose dramatically and famine and disease in epidemic proportions then gripped many parts of the country, causing massive life loss, especially among the indigenous populations (Orozco y Berra, 1938; Berthe, 1970; Cope, 1994).

Frosts repeatedly affected the sugar cane plantations in Oaxaca during this decade. Among the more successful of the sugar cane ventures in the region were the six plots of cane grown on the haciendas of San Bartolomé in the mountains to the east of the Valley of Oaxaca. A document from 1834 dealing with transactions and deeds of sale of the hacienda details the ruin that befell one of its sugar plantations named San Nicolas, by a series of severe frosts throughout the decade. Frost destroyed the entire harvest in 1694, while the effect of a further episode of severe frost on the recently sown sugar crops only three years later in July 1697 was completely devastating, leading to a total abandonment of the venture and the sale of the property.[13] In the north, meanwhile, the problems of drought and harvest failure were followed by epidemics (Deeds, 2003: 196) and other natural disasters, including an earthquake in 1696 (Neumann, 1969; Aboites Aguilar, 1994), a combination of circumstances which may go some way to explaining the apparent increase in raiding by nomadic groups throughout this decade (Griffen, 1979: 20; Martin, 1996: 439).

There were other more widespread problems of banditry, theft, assaults and a wide variety of criminal actions recorded across the country at this time (Florescano, 1972). Indeed, the nationwide calamities reported in the 1690s might in part have contributed to an increase in crime generally, which Florescano has referred to as 'the third plague', after harvest crisis and epidemic disease. The urban riots recorded in Guanajuato and Mexico City in 1692 can be considered in this light.[14] Either way, various strategies were deployed by the Viceroy in an attempt to reduce the scale of the problem. In 1699, for example, branding on the foot or the hand was the threat levied at anyone under accusation of theft, while removal to the Yucatan Peninsula was another of the penalties. The Viceroy did seek permission to apply corporal punishment in 1706, though this was overturned by a superior tribunal. A period of crisis between 1709 and 1710, however, witnessed an acceleration in criminality and once again renewed calls for more punitive measures, which were finally implemented in 1719 when authority was given to apprehend 'all... delinquents, robbers and highwaymen in areas that are populated and those that are not' and to sentence them accordingly.[15]

'Cattle dying in the field': drought and pestilence in the mid-eighteenth century

A period of crisis is also recorded in all three regions around the middle of the eighteenth century. Droughts across the Bajío had contributed to harvest shortfalls in September 1749; by early November that year food shortages were recorded in some parts of the region. Prices soared to unprecedented levels (48 and 52 reales per fanega) and food riots are again reported as having taken place.[16] Dramatic loss of harvests and the death of livestock left many individuals without any means of survival and with no alternative but to sell off their possessions, property and land. Indigenous residents of Léon (Guanajuato), Juan Santos Ramirez, Martin Fe la Cruz and Luisa and

José Ramirez were among those with no option but to put their land and farm buildings on the market 'due to the terrible dryness of the [present] year'.[17] All sections of society appear to have been affected, even the wealthy individuals. At the height of the crisis, for example, Christobal Manuel, described as a 'rich latino resident of the juristiction of Léon', stated he had been forced to sell off part of his property 'because of the sterility of the year' which had left him 'impeded because of the lack of oxen with which to sow'. His family were apparently 'suffering terrible neediness because of the lack of maize which because of the scarcity has such an inflated price'.[18]

Drought and harvest problems are also documented elsewhere in the country around the middle of the eighteenth century. Drought induced harvest failure might have also been responsible for the reported food scarcities in Oaxaca between 1746 and 1748,[19] when grain stores and granaries were depleted, prices rose and the local administration in Antequera (Oaxaca City) was requested to provide food or financial aid for the poorer sectors of society who faced starvation.[20] Yet it is clear that drought and food scarcity appear to have particularly affected people in the more arid north of the country at this time, possibly accounting for the resurgence in social unrest and physical violence in the region. As demonstrated in chapters 4 and 5, repeated droughts in the years between 1749 and 1752, in 1755 and again in the later 1750s had serious implications for pasture availability,[21] coincided with a period of livestock pestilence,[22] and thus both directly and indirectly resulted in the death of livestock.[23] There were shortages of grains (especially maize) and populations faced starvation,[24] but because of the possibility of cross-contamination it was felt that dead livestock had to be disposed of and not sold for meat.[25] Nonetheless, illnesses recorded among some of the more poverty stricken members of society at this time were attributed to the consumption of infected carcasses.[26] As the problems persisted in the north of the country throughout the 1750s and early 1760s,[27] epidemics and pestilence took hold,[28] which in turn led to an abandonment of some of the mines in this northern region and to the migration of hundreds of people. The scale of the crisis resulted in special requests from local communities for financial donations to create a public pharmacy to allow people to access essential medicines.[29] It seems that this request went unanswered, however, as there were renewed claims that there was still a need for a public pharmacy forty years later in August, 1792.[30] Epidemics were also recorded in other parts of the country at this time (Cavo, 1949), some of which were attributed to the desperate consumption of poor quality famine foods by the most poverty stricken, starving members of society.[31] One must also bear in mind that increased mobility of people at this time might have also facilitated contagion. Though the period of disease might not be related directly to the drought conditions of the 1750s per se, food and water scarcities inevitably resulted in a much weakened and hence more susceptible population.

Both abnormal weather and epidemic disease could lead to harvest shortfalls, in the latter case by effectively reducing the availability of agricultural labour, and it is difficult to identify from archival sources alone whether harvest problems were a function of weather or disease. The epidemic which was recorded in Chihuahua in 1764, for example,[32] may well have been associated with the well-documented 'common and great epidemic',[33] consisting of both typhus and smallpox, which swept across Mexico between 1763 and 1764 (Cooper, 1965). However, the harvests in Chihuahua had also been affected by the 'rigorous drought' between 1764 and 1766,[34] with hailstorms in 1764[35] (considered in chapter 5) perhaps accounting for the limited supplies of grain arriving in the alhóndiga (as noted in chapter 4). In this case, it seems that unusual weather events might have exacerbated the difficulties faced by a society already rendered vulnerable by epidemic disease.

Drought related famine thus prevailed in central Mexico for the early years of the 1750s and problems of food scarcity are reported in the south of the country immediately before the 1750s. The period between 1751 and 1766 appears to have been one of protracted climate related crisis in the north of the country. This supports tree ring evidence of a prolonged drought in the middle of the century, which is now thought to have been the second largest drought for the period between 1647 and 1992 (Dias et al., 2002).[36]

'Extraordinary heat, rigorous droughts and shattering hurricanes': crisis in the 1780s and 1790s

Health observations recorded in the early 1780s highlight an awareness of changing climatic conditions and the associated implications at the time. The *Gazeta de México* of 10 March 1784 reported how, from 7 December the previous year, 'evil winds from the south and southeast have blown with greater frequency'. This was associated with the reappearance of 'sicknesses', including 'pneumonia, pleurisy, apoplexy, angina, inflammation of the throat; and finally, other ailments, which although customary during this season of the year, began to be felt with greater frequency and virulence'.[37] Moreover, it seems that the Spanish Crown might have anticipated the possibility of an imminent crisis around this time. On 10 May 1784, in what can now be considered something of a portentous act, a royal order had been issued in Spain requesting that 'all the heads of the Indies send each six months' notice of the weather experienced in these dominions: noting whether the rains had been scarce or abundant and noting also the nature of the harvests of fruits and other produce'.[38] It is not clear whether this order came in response to a fear of more widespread harvest problems or a general growing awareness of the potential for climate induced agrarian problems in the Spanish American colonies as a whole. However, the order preceded one of the most widespread and devastating famines in Mexican colonial history.

The consequences were particularly devastating in the agrarian heartland of central Mexico. The first signs of a problem in this region may have actually been recorded in documents from 1782, when a local councillor from Léon reported that 'the population is suffering from... scarcity and a need for seeds of maize and other crops'.[39] By early 1784 the manager of the grain store in Léon recorded 'a lack of maize ... because of the scarcity of the harvest' of that year.[40] Scarcity in turn forced prices up and there was concern that this would prevent the poorest sectors of society from accessing the most basic foodstuffs.

By 1785 problems of scarcity were becoming more widespread. A circular sent by the Viceroy, Conde de Galvez of Léon in 1785, for example, suggests that food shortages were 'beginning to be experienced in most of the kingdom because of the lack of rains and in anticipation of the winter'.[41] Actual grain shortages and speculation meant that prices sky-rocketed across the Bajío to 48 reales per fanega and hunger was rife across the north of the country. Rogation ceremonies were held in Chihuahua and Mexico City and appeals were made for divine intervention to alleviate the suffering.[42]

Though irrigation may not have always been that effective in alleviating drought induced harvest losses during particularly severe droughts (Yates, 1981), the expansion of land under irrigation remained a key coping strategy during times of drought related food scarcity. In 1785 communities across central Mexico were encouraged to grow more crops under irrigation, though in some cases this necessitated financial assistance. A letter to the *Gazeta de México* from the Bishop of Michoacán, dated 19 November 1785, suggested that, 'to facilitate more and more sowing of irrigated maize it has been determined that one thousand pesos be distributed in this parish among the poor, well off Spaniards, Indios or Mulattos who wish to sow'.[43] People from all sections of society were being encouraged to grow irrigated maize in stretches of territory across central Mexico that were not normally considered suitable for this particular crop, including the so-called hot lowlands (tierra caliente) and cold uplands (tierra fria) in the spring of 1786.[44] In part this could be because, as one document suggests, there was 'more water for irrigation of maize' in many of these districts at this time.[45]

Elsewhere, the same period of crisis led to more desperate measures and the issuing of decrees instructing some sectors of society to engage in agricultural investments for the common good. The local authority in Chihuahua released a circular on 11 October 1785 suggesting that the Indio labourers and other workers should be persuaded not to abandon their pueblos and haciendas, but rather to 'provide their personal services to the cultivation of the fields'. Moreover, 'being a time of calamity', the indigenous communities 'were obliged to work in occupations considered useful to the Spanish, Mestizos, Mulattos and other Castas and vagabonds and thieves ought to assist in the applications of these people to the labours of the field'. In Chihuahua this was a time when the town's jail was overflowing with individuals accused of collaborating in the hostile indigenous attacks against the colonial settlements. Employing the inmates in agricultural efforts thus offered a potential solution to overcrowding in the local prisons as well as providing some hopes of a better harvest the following year.[46]

As had been the case with the crisis of the 1750s, some people had little choice but to sell part or all of their property, usually far below the true sale value, as desperate vendors were keen to point out. Documents from Leon chart the sale of several houses 'for more or less half of the just value and price',[47] and highlight the 'the extreme need for help because of the hunger'.[48] Some benefactors in Chihuahua donated money to aid the poor in 1785,[49] and local councils of Guanajuato also began to make charitable donations,[50] and circulated the names of wealthy individuals in Leon who could provide food or financial aid.[51] A document from Léon compiled on 15 February 1786 similarly details the donation 'by hand' of 50 pesos from one Ignacio Gomez Ploeo, resident of Léon, to help the 'poor, demented' people in the city at that time.[52] Local councils of Guanajuato also made charitable donations the following month because of 'the scarcity of maize and other grains',[53] and took the unusual step of circulating the names of wealthy individuals in Léon who could provide food or financial aid should the need arise.[54] The Superior Government also granted remission of tribute and sales tax on foodstuffs for the mining workforce in the Bajío (and Zacatecas), but the heartland of agrarian production, the area that had so regularly provided foodstuffs for other regions during periods of calamity, was especially hard hit by this subsistence crisis. Food scarcity led to a fall in silver production and so royal revenues dropped (Hamnett, 2002: 111). The economic well-being of the region and the colony as a whole was affected. In Chihuahua, meanwhile, after the long drought, the 'continuous rains' in August 1786 brought little relief. In fact, it transpired that the rains actually damaged the seedlings, so that hopes of a more successful late harvest were dashed,[55] and the repercussions were described as 'deplorable' for agriculture, but also, by extension, for other sectors of the economy in the region.[56]

Several years of hardship would follow. Three years after the first signs of drought, 300,000 people had died and others had been incapacitated and were thus in no fit state to cultivate lands when the rains did arrive. There are reports of disease epidemics in various parts of the country during and following the crisis.[57] Between 1785 and 1786 burial registers in Guanajuato alone recorded 16,000 deaths.[58] The droughts had served to diminish the amount of seed stock available with which to sow (Florescano, 1981) and tribute demands remained impossible to satisfy in the wake of the prolonged crop shortages.[59] Five years later, some towns in the Bajío were still struggling to meet these demands. In 1790, for example, the communities of San Miguel, Cocuillo and San Francisco y la Concepción de la Rincon were still reeling from the 'inconceivable losses' of the 1785–6 famine.[60]

Although drought and harvest crisis are recorded in Oaxaca at this time, indigenous land retention, the strength of the local cochineal trade, and less of a reliance on wealthy landowners for food provision in this region may have given a large proportion of the population somewhat greater independence and, it follows, more subsistence capability relative to other regions. As seen in chapter 4, some communities in Oaxaca were even able to produce a surplus for those regions that had suffered losses. Yet although some communities in Oaxaca may have fared better than most during the 'Year of Hunger', they were not all spared the impacts of freak weather events and other natural hazards during the 1780s. The decade had in fact started badly. In 1781 the 'cruel epidemic of smallpox' which had ravaged much of the country the previous two years was taking devastating effect in Oaxaca. There was also 'terrible hunger' recorded in many districts in this year, though possibly a function of the lack of agricultural labour as a result of the epidemic rather than unusual weather.[61] Two years later on 15 December 1783, a series of worrying phenomena and unusual 'storm' events were recorded in Teotitlán del Valle, just as the 'evil winds' recorded in the *Gazeta de México* had begun to blow.[62] 'A great snowfall covered the fields of the area … extremely heavy rainstorms, which were no less damaging to the sown crops than the strong drought of the previous months … the streets were transformed into rivers and the houses were left flooded out'. The events were apparently compounded by 'extraordinary heat, rigorous droughts, shattering hurricanes, frightening falls of hail and storms accompanied by prolonged underground noises' (Gay, 1950: 427–428). Only a few months earlier, volcanic ash clouds were observed and ash falls had covered fields and roads across the region, seriously disrupting agricultural production.[63]

Other non-climatic natural disasters compounded the problems faced in this decade. 'Tremors of the ground' were considered 'very frequent' phenomena in Oaxaca.[64] Gay, for example, has suggested that earthquakes were recorded across the region in 1603, 1604, 1608, 1682, 1696, 1727, 1787 and 1789. Of these, the event of 1787, which was experienced across southern Mexico but particularly in Oaxaca, appears to have been among the most terrifying. Various reports suggest that the tremors that year began either on 22 or 24 March, but all accounts agree that they were most strongly felt from Wednesday 28 March at around 11 o'clock in the morning until 9 April.[65] There was considerable damage to houses and administrative buildings, including the jail and the church in Antequera, while those buildings constructed of adobe completely collapsed.[66] As Gay notes, the earthquake of 28 March was also felt as far west as Acapulco. His description of the event highlights that there might have been other associated events, not least a possible tsunami. 'The sea was seen to retreat and then return, and later rise up and drown the harbour/bay area, repeating this sequence various times in the space of 24 hours.' Houses were destroyed and 'great quantities' of livestock were washed up on the shoreline. All the while there were 'frequent earthquakes', possibly aftershocks, reported across the south of the country. Nor was this an isolated incident. On 3 April of that year at 9 o'clock in the morning there was another powerful earthquake, while others still were reported in the following two years (Gay, 1950). According to one document, 'on the 5, 7 and 9 of May in the year 1788, there were felt in this city two movements of the ground'. The same month there had also been a series of extreme climatic phenomena, including a possible hurricane, recorded in Antequera. 'The heat had been great, and there had not been more than two rainshowers in the whole of the previous month ... on the 9 May a rainstorm of some consideration took place and a little later a very strong hurricane and lightning storm, with fork lightning'. There was considerable damage. The dome/roof of the local church was wrecked, though fortunately only two deaths were registered.[67] Other references suggest that these unusual weather events may have had impacts across a broader region. According to a note from the Alcade Mayor of Mezitlán compiled on 31 December 1788,[68] people of the Indio community of the town of San Pedro on the edge of the cabacera of Mezitlán were apparently prevented from sowing their fields because of floods. Yet overall, 'thanks to divine mercy', relatively good harvests were recorded across the province in this year.[69]

The early years of the 1790s saw a return to subsistence problems in some parts of Oaxaca. In San Andres Zimatlán, for example, the harvests of 1793 were described as 'poor'. Prices for maize and wheat increased accordingly,[70] and repartimiento payments apparently became impossible to fulfil.[71] Reports indicate that the maize harvests in Teococuilco in 1794 and 1795 had been similarly 'scarce' and that only half the normal quantity of wheat had been reaped. There was a shortage of grain recorded across the whole of the province generally, although there was some regional variation. The town of Santiago Comaltepeqye was able to secure a relatively good harvest,[72] and there were hopes of a better harvest in the town of San Pablo Guelatao.[73] In 1794 the indigenous community of San Jeronimo Sagache highlighted the apparent cause of these harvest problems, arguing that the 'scarcity of rains experienced in this year has had an unfortunate effect on the harvest', leaving them unable to meet their repartimiento payments. Such difficulties led them to ask the Alcade Mayor whether it would be possible to delay payments for a period of 12 years.[74]

Worse events were to descend on the region with the return of smallpox in 1795. Various attempts were made to contain the illness. Within Oaxaca, 'the infected were separated in the first instance and were sent to a site set aside where they could be cured in a hospital'.[75] Letters were sent to other Intendencias asking for them to co-operate in the attempt to stop the spread of the disease.[76] A special 'vaccination house' had even been established in Querétaro.[77] Action was thus taken to avoid the contagion spreading, but by 1797 the illness had diffused across much of the country (Cooper, 1965).

The 'afflicting spectacle of famine' at the turn of the nineteenth century

The year 1799 was, according to Swan's (1981) data, generally very cold in Mexico. In some places in central Mexico frosts were reported in August and snow in December. Hail also threatened both crops and livestock (Beveridge, 1921, 1922; Florescano, 1972). Portillo (1910) reports on a series of poor harvests in 1801 and 1802 brought on in part by a plague of locusts which invaded 'the Spanish colonies in south and then central America, affecting the Mexican realm in 1802'. Communities were galvanized to act collectively to reduce the likelihood of an impending crisis. Notices were issued offering 'ten pesos for each arroba of locusts that were collected and five pesos for the collection of the larvae which resulted from their procreation' in an attempt to alleviate the problem. Nevertheless, by 1804 there was a great scarcity of grain (particularly maize) recorded in some regions of the country, and particularly in Oaxaca, where people were forced to 'roam through the fields looking for edible roots and weeds so that they would not die of hunger' (Portillo, 1910).

Various different forms of anomalous weather and epidemic disasters are also recorded elsewhere across the country between 1801 and 1804. There are reports of repeated flood events in Oaxaca at this time.[78] In Iztlahuaca, 'excessive increases' in rainfall and consequent flooding were recorded in 1802,[79] while as shown there was a series of consecutive flood events thought to be on the Rio Mixteco in the north of the region, recorded on 11, 12 and 13 January 1804.[80] In central Mexico the River Lerma was apparently 'higher than the roads and in the houses' during the wet season of 1804,[81] while heavy rains and flooding had been documented in the northeast of Mexico (specifically, Coahuila, Nuevo León and along the Lower Rio Grande) in June 1802, and the harvests failed in Chihuahua following 'extraordinary drought' conditions in 1804. These unusual sequential weather conditions are thought to have led to a regional economic crisis and to have contributed to the emergence of disease in epidemic proportions in the north of Mexico (Butzer, 2003). Grain scarcities were apparently localized, but there was concern in August of 1804 that food supplies, in the city of Chihuahua at least, would remain a problem until at least the following month. In fact, food shortages were recorded in the region between 1804 and 1806.[82]

It was this weather and famine ravaged and pestilential landscape which polymath scholar Baron von Humboldt described after a visit to New Spain between 1803 and 1804. Reflecting in his 'Geographical description of New Spain' on the levels of poverty, malnutrition, hunger and disease that he had encountered in central Mexico in particular, he suggested that a lack of long-term planning for food crises explained the prevalence of disease among Mexican communities. 'In a country where the naturales [indigenous people] live from day to day', he suggests 'the town suffers immensely so the naturales are kept alive with fruits and roots and cacti and this bad food produces illnesses and in general the scarcity observed is accompanied by mortality among children' (Humboldt, 1811). A compounding problem lay in the fact that the majority of the population was dependent upon maize production. 'The maize is the principal foodstuff of the town as well as for all domestic animals.' This was not so much a problem in normal years or if food production had kept pace with the population expansion that characterized eighteenth-century Mexico but, for Humboldt, 'the want of proportion between the progress of population and the increase of food from cultivation renews the afflicting spectacle of famine whenever a great drought or any other local cause has damaged the crop of maize' (Humboldt, 1811: 175). Drought or extreme and unexpected events could thus act as 'triggers' to agrarian crises where insufficient agrarian interest, investment and production were combined with population expansion. Moreover, this kind of crisis could have repercussions for other sectors of the economy. 'When the harvest is bad, if there is a lack of water, or early frosts, the scarcity in general has calamitous effects. The chickens, the turkeys and even the cattle are affected by it' (cited in Dunn, 1988: 98).

From Humboldt's perspective, therefore, investment in mineral activities at the expense of agriculture, a focus on subsistence-based production and the reliance on maize as a principal commodity, coupled with a highly interconnected economy, most certainly rendered society much more vulnerable to environmental forces such as unusual or extreme weather. In fact, only a few years later, a prolonged three-year drought would test the resilience of Mexican society and highlight some of the problems that Humboldt raised. 'Scarcity of rains' between 1808 and 1810 triggered a devastating, widespread agrarian crisis. By this period, population expansion and a growing resource monopolization by an emerging elite across many parts of the country had effectively created a population that was far more vulnerable to the impacts of such unexpected climate changes. Recognition of this vulnerability is thought to have helped fuel the agrarian unrest in the heartland of the Bajío which would eventually sweep across the country in the drive for Independence (Tutino, 1986).

From Crisis to Insurrection: Vulnerability and Popular Unrest in the Early Nineteenth Century

As suggested in the previous chapter, although colonial society in Mexico was litigious and there were various forms of dissension, violent and collective action was unusual. In September 1810, however, there was in Van Young's (1986: 385) terms, 'a sudden flash of violent rebellion' across the country. Unusual in terms of its scale, the numbers of people involved, the geographical extent of their involvement and the duration of the popular participation, this was a period of 'preternatural violence' (Van Young, 1988). Reflecting months and even years of conspiracy and unrest, particularly in the central regions of the country,[83] there began a process of rebellion that would attempt to overthrow the Spanish regime. Although by 1816 much of the country had been pacified, small pockets of rebellion prevailed and would help constitute the more successful broader expression of popular discontent manifest in the Wars of Independence from Spain between this time and 1821 (Van Young, 1986).

The seeds of this insurrection are manifold, complex and diverse. Hamnett (2002: 75), for example, suggests that social unrest was a function of a variety of structural problems and social, economic, political and also environmental circumstances. In the late eighteenth and early nineteenth centuries, the Spanish state was in crisis and was facing demoralization, chaos and collapse (Herr, 1958; Hamnett, 1985, 1997). Napoleon's invasion of the Iberian Peninsula and the forced abdications of Charles IV and Ferdinand VII in 1808 threatened the unity of the Spanish empire. Spanish insurrections against the French at the time and the final defeat of the Bourbon regime in 1808 may have helped spur popular movements in America. By the second half of the eighteenth century, the colonial order was beginning to fracture (Lynch, 1992: 114). The drives toward market integration and the growth in international trade which stimulated a phase of remarkably rapid change, reform and globalization had exacerbated a growing and 'formidable inequality' (Ponzio, 2005: 437). Some groups benefited from the economic developments of the 1700s, while others suffered severe economic deprivation (Andrews, 1985). As seen, a series of climate and environmental crises may have first revealed and then compounded these problems.

Among those sectors most severely affected were the cochineal producers of Oaxaca. Indeed, the implications of the complex array of social, economic, political and environmental problems faced toward the late eighteenth and early nineteenth centuries is perhaps best illustrated through a focused investigation of the dynamics of the cochineal industry at this time. After a period of economic fluorescence in the 1700s, the cochineal industry in Oaxaca fell into crisis towards the last decade of the eighteenth century and the first ten years of the nineteenth century, despite continued high demand for the product in Europe and beyond. There was a dramatic reduction in cochineal production from 1786 onwards, from around 30,000 arrobas a year to less than 6,000. In fact, the production of grana fina more or less halved between 1758 and 1820 (Baskes, 2000: 54).

A commercial survey produced in the early nineteenth century proposes a number of explanations for this economic downturn. The Real Ordenanza de Intendentes introduced as part of the Bourbon reforms might have had a particularly dramatic effect on cochineal production. Article 12 saw the abolition of the repartimiento trading system and, as noted, the removal and replacement of Alcades Mayores. This served to withdraw an important source of Indio credit. The Alcade Mayores were no longer in place to maintain the trading relationships and credit systems that had been established, and without the ability to collect debts cheaply and safely, few private merchants stepped in to take on this role (Baskes, 2005). The establishment of the Intendencia and the removal of the Alcades Mayores also contributed to a fall in the number of Indio towns of the Bishopric of Oaxaca and 'this was one of the most immediate causes of the decline' in cochineal output. The survey argues that Indios had increasingly begun to abandon towns to avoid tribute and repartimiento deliveries of cochineal harvests, preferring instead to work in the city as traders, bricklayers and construction workers. This caused particular problems because the 'Castas and Españoles that had been dedicated to sowing fields lacked the scientific knowledge and dedication' to produce cochineal, while 'the commercial traders were interested in making profit from the grain'. Certainly, most attempts by the Spanish to cultivate the insects had met with failure.[84] There were also concerns that the Indios responsible for cochineal production had 'fallen prey to drunkenness, idleness and laziness'.

Other explanations for the decline were also forwarded, including 'the problems of the times, inclemency, alterations in the elements, natural deterioration ... in all the fruits of the earth, disease among men and cattle, war and all the other general events that had caused a decline in cochineal and an increase in agriculture'. In June 1779 Spain had joined France in the War of American Independence and this may have had some repercussions for cochineal production. Some Alcades Mayores at this time attempted to reduce the fixed loans or credit advances during this period, though this met with fierce resistance from the indigenous producers and in one case, the Real Audiencia (Mexico's high court).[85] Yet such instances of variation from the fixed loan rate provided by Alcades while they still existed were rare.

Climatic variability was thought to have played a vitally important role in changing levels of production. Cochineal required 'a special climate, benign, nothing extreme … a humid temperate climate that is found in the whole province'. Unusual or unexpected weather events could destroy the harvest (Donkin, 1977), so making repayment of loans and credit impossible (Baskes, 2005). It was recognized that a lack of rain had presented problems in some years, when the city of Antequera and the majority of the Bishopric of Oaxaca and the sown fields had suffered. Moreover, as highlighted earlier, there had been particular climatic problems around the turn of the nineteenth century when 'strong rainstorms and hail' which had covered the nopal cacti reduced the harvest of the insects and resulted in an 'inferior' output of cochineal. In fact, in the years leading up to 1810, it was noted that there had been 'scarcity of harvests due to bad weather'.[86] It was thus understood that a range of environmental, social, economic and political factors coalesced to devastate Oaxaca's most important export commodity by the early nineteenth century. The once buoyant industry, though still producing cochineal for export, witnessed a dramatic decline on the eve of Independence.

Of course, it was not just the cochineal industry that faced crisis in the late eighteenth and early nineteenth centuries. At the same time, there were mounting tensions between rich mine operators and poor mine workers. The latter generally enjoyed the customary right of partido, whereby labourers were paid partly in cash and partly in a share of the ore. Any changes to the partido resulted in riots among the mining communities. This was the case in Guanajuato in 1766 following threats to completely abolish partido arrangements (Hamnett, 2002: 96), on top of fiscal pressures, the tightening up of tribute collection and imposition of excise duty on the locally produced alcohol. This riot served its purpose. Indeed, on this occasion Viceroys Cruillas and Criox relented, though the rioters were severely punished. Some 3,000 people were brought to trial, 117 ordered to leave their homes, 73 subject to whippings and 35 executed (Hamnett, 2002: 98). Nonetheless, further threats to the partido system the following year, combined with the decision to undertake a new census and the religious grievances over the expulsion of the Jesuits, similarly resulted in unrest among the mining communities. This time, hostility was directed specifically against the Alcade Mayor.[87]

There were parallels in terms of the loss of customary rights in the agricultural sector. By the close of the eighteenth century, a growing resource monopolization by an emerging landed elite across many parts of the country, but particularly in the Bajío, had effectively created a more marginal – and as Tutino (1986) has pointed out – an increasingly dependent underclass. Cereal cultivation had brought prosperity to the Bajío. Haciendas were able to exercise control and exert pressure on an abundant labour force, increasing rents and replacing customary rights with cash payments. Evictions were common-place and some people were forced to migrate (Tutino, 1986). Criminality, vagrancy and small infractions in the cities increased with the periodic influx of rural migrants to the urban centres, seeking safety and relief, and the number of village riots in colonial Mexico also 'mushroomed' around this time (Hamnett, 2002: 88). Riots were recorded in Etla on a number of occasions in the 1750s and again in the 1790s.[88] To some extent these manifestations of unrest reflect more structural factors. Between 1670 and 1810 the population of the region rose significantly, which placed pressure on unevenly distributed resources (Taylor, 1972: 34). There was also considerable tension between village cultivators and private estate owners, metered out in the many lawsuits over land and water (Tutino, 1986). The records of land disputes, for exam-ple, highlight the insufficiency of existing land resources that were no longer able to sustain an expanded population. Many of the lawsuits raised by indig-enous communities at this time highlight the erosion of the 600 vara legal fund through hacienda encroachment (Endfield, 1998; Hamnett, 2002: 88). These factors were common sources of legal (and, more rarely, physical) conflict across broad geographical areas of the country by the close of the eighteenth century. A number of areas had become recognized as trouble spots of tension and unrest, including, as Tutino (1986) notes, communities across Guanajuato. Such problems, however, were not necessarily widespread and did not in themselves give rise to insurrection.

Subsistence crisis between 1808 and 1810 created common conditions of hunger and unemployment across many parts of New Spain (Padilla et al., 1980; Conteras, 1999; Servin, 2005). There was 'a lack of maize, beans and other seeds of basic necessity' resulting in 'want, hunger and calamity' across the country. Indeed, 30 out of the 41 districts across Mexico experienced drought and crop losses. In Guanajuato alone, the first effects of the crisis included the loss of half of the sown maize (Gomez and Armando, 1995; Hamnett, 2002: 119), while there were shortages of bread recorded in the capital city.[89] Prices soared and shortages of food affected mining activities, while the death of livestock served to aggravate the cereal crisis by reducing the number of beasts of burden available for work in the field. Supplies of raw materials could not be transported to domestic pro-ducers and manufacturing slowed down. There was also a substantial increase in the price of meat. Indio communities found themselves unable to fulfil tribute demands and they were plunged into debt. By February 1809 some indigenous communities in Léon owed $6,183 in unpaid trib-ute.[90] These difficulties were compounded by shortages in the supply of mercury, which as noted was critical to the amalgamation process of silver extraction and so pivotal for the smooth running of mining operations. Overall revenues, therefore, decreased. In the north of the country too, the 'scarcity of rains' in 1809 'produced a depleted harvest of maize in these territories', though some areas had been particularly dramatically affected. There was, for example, 'an urgent need for food and extreme deprivation' recorded in Chihuahua City. As with previous periods of harvest shortfalls and grain scarcity, attempts were made to secure supplies of maize from outside the area to relieve the problem.[91] Trade and commerce, however, had been severely disrupted by the political and social problems emerging across the country,[92] and by 26 April 1810 there was 'grain scarcity due to the lack of rains experienced in the entire virreynato'.[93]

The dearth also produced population dislocation. This, perhaps more than the subsistence problems per se, would have serious consequences for political and social stability. This was a factor recognized by the colonial government, which in 1809, fearing social unrest during the peak of the crisis, issued a decree attempting to curtail population mobility. There was particular concern that 'as was common during years of calamity, the poorest sectors of society, and especially the Indios' would 'abandon their houses, leave the places and towns of residence, causing problems'.[94] There had been a general lowering of living standards in the fifty or so years preceding the insurrection of 1810 and there was a general discontent among Spanish American professional classes at the lack of opportunity. By the close of the eighteenth century these factors, coupled with the recovery of village population levels, provided an additional dimension to the sources of local conflict. Circumstances in the Bajío in particular provided a 'fertile ground for rebellious sentiments', and, as the agrarian heartland of the colonial political economy, it is no coincidence that this region represented the hearth of insurgent activity under the leadership of the charismatic Hidalgo.[95]

There was a strong rurality to the rebellions of the early nineteenth century (Van Young, 1988). The colonial regime was threatened by armed insurrection specifically in the countryside. There was a general absence of any large-scale urban uprising, which Van Young has argued can be explained by a variety of social and economic conditions which militated against concerted forms of political action. With some exceptions, facilities and services in the urban centres may well have helped to support the most disadvantaged members of the populace. Via the positos and alhóndigas, the city governments attempted to ensure a good supply of foodstuffs. Yet such services were not always so accessible to rural communities. Pro-royalist, anti-rebel propaganda was strongest in the cities, while social control and repression also put paid to any unrest. Indeed, the heightened security measures in the cities, together with the presence of permanent military garrisons in some places, relative to the rural areas, effectively discouraged urban uprisings on the whole, a fact borne out by the limited number of urban riots recorded during the colonial period. Indeed, most Mexican cities lacked a tradition of rebellion. Moreover, the Church actively sought to preach against rebellion.

In the rural areas of New Spain, however, struggles over landownership, labour services, the erosion of rights, including land and water rights, and status, and challenges to customary practices and traditional ways of life, combined to push communities 'to the brink of despair' (Hamnett, 2002: 75). Wages were low relative to city dwellers and in the central region of the country at least, where the rebellion was fomented, the effects of meteorological events were more dramatically and often, as shown, tragically, experienced by the marginalized underclass.

There had been previous subsistence crises of similar or even greater magnitude. The events which culminated in the Year of Hunger, for example, remain unmatched in terms of the scale of life and economic losses, yet, perhaps paradoxically, this period did not stimulate widespread popular and collective unrest. This, of course, may be because of the sheer scale of disruption, disease and death. By the turn of the nineteenth century, however, a suite of external and local, natural and anthropogenic factors coalesced in a unique way regionally across Mexico, but also in the Spanish empire as a whole. Population expansion, gross inequalities in ownership and privileges of access to natural resources, loss of customary rights and the general erosion and undermining of traditional ways of life were common problems. The Bourbon Reforms had served to undermine or threaten economic well-being for some sectors of society and there were mounting grievances as a result. As Hamnett (2002: 107) suggests, it was this coming together of 'disparate agencies' that resulted in widespread insurrection. Against this backdrop of social, political and economic problems and with the impacts of previous crises still very much at the forefront of public consciousness, the subsistence crisis of 1808–10 could be seen as a trigger of popular unrest.

Chapter Seven

Regional, National and Global Dimensions of Vulnerability and Crisis in Colonial Mexico

Introduction

The rich colonial archives of Mexico reveal that seasonal climate variability, extreme events and related impacts have posed a dynamic set of problems and opportunities for society across the country, altering the context of vulnerability, adaptability and indeed survival. Seed, grain and food storage and trading, the marshalling of social networks, irrigation and associated water management and storage, represented important pre- and post-Hispanic strategies geared towards reducing vulnerability to climate change and its impacts at a range of scales. Indeed, such adaptive strategies to some extent appear to have been relatively successful in dealing with single year or predicted events. Lower frequency climatic variations with larger amplitudes, however, and unexpected changes in weather acting on an unprepared or already vulnerable society, tested social resilience and the capacity for human adaptability. The unique social, economic, environmental and climatic characteristics of each of the case study regions, however, were pivotal to influencing the relative vulnerability, resilience and adaptability of different sectors of society to climatic variability at different points in time. These conditions necessarily determined the way in which the effects of normal seasonal climate changes as well as those associated with more extreme or unusual weather, including those events which had widespread impacts, were experienced and perceived, understood and addressed.

Prolonged Drought and the Conditions of Crisis in Late Colonial Chihuahua

Records pertaining to Chihuahua indicate that drought has played a key role in the region's socioeconomic development and has represented a constant problem for both pre- and post-Conquest society. The majority of drought induced harvest crises recorded in the colonial archives date from the seventeenth and eighteenth centuries. By this stage, in this hitherto little settled region, there had been an expansion in the number of garrison towns and presidios, as well as an increase in their populations, in conjunction with the development of mining and agricultural activities across the region, meaning that there was a greater number of potentially vulnerable people and an increased risk of impact associated with drought events. Evidence drawn from primary archival sources consulted in this project not only supports suggestions that the region experienced a number of multiple year droughts during the colonial period, but has also indicated that some years within these prolonged drought 'phases' can be highlighted as being particularly problematic. Archival records suggest that repeated droughts in 1690–3, 1724–7, 1739–41, 1748–52, 1758, 1760–5, 1770–3, 1785–6, 1804–6 and 1812–14 contributed to harvest failure and famine, affected the health of both human and animal populations, disrupted mining activities and may have also triggered or at least contributed to periods of social unrest.

A number of droughts in Chihuahua – notably those of the 1720s, 1750s, 1770–3, the 1780s, 1804–6 and 1808–10 – appear to have had wholesale impacts which affected all sectors of society. The repercussions of other periods of drought, however, or drought related harvest problems, may have selectively affected certain sectors of the population and the regional economy. Prices for grains in particular, but also for meat, increased following drought. Not all price fluctuations and indeed, scarcities, identified in the archival sources were necessarily climatically driven. Speculation has of course been recognized elsewhere in Mexico, including the Bajío region, but was clearly a problem in eighteenth-century Chihuahua. Individual merchants and landowners often deliberately purchased in bulk, sometimes outside of the region, to sell at elevated prices in the cities of Chihuahua, or stored and withheld grains to simulate scarcity of the most essential commodities. Either way, price rises, either legitimate or deliberately forced, worked against the poorer indigenous classes.

It seems that it was the same groups of people that disproportionately succumbed to illness and disease epidemics during or following drought induced food scarcity and famine. Some of the recorded epidemics in Chihuahua, specifically in 1779–80 and 1813, relate to nationwide epidemics that are discussed in more detail elsewhere (Cooper, 1965). There may, however, also be strong links between drought, harvest crisis and disease in this region, with disease appearing sometimes in epidemic proportions, and among animal as well as human populations. Although it is open to conjecture whether disease epidemics can or should be attributed to weather conditions and food scarcity or also to the socioeconomic organization of the society, authors of some of the archival documents consulted do draw direct links between drought and disease in 1752, 1758, 1764, 1785 and 1809. Drought interspersed with heavy rainfall may have at least triggered deterioration in ecological and socioeconomic conditions which may have served to magnify the impact of infectious disease on these occasions.

Archival records suggest that many of the instances of violent incursion and attacks during the seventeenth and eighteenth centuries may similarly correspond to periods of recognized drought and some informants again actually imply that attacks were directly associated with the effects of drought. Pressure on limited resources and the difficulties of leading a subsistence and hunting/gathering existence in this region may have been exacerbated during successive drought periods and thus could have helped stimulate at least some of the raids. Production and distribution of basic and emergency supplies might have been disrupted by such attacks, compounding the problems of food scarcity during periods of subsistence crisis. The implications of prolonged or successive drought in colonial Chihuahua, therefore, were myriad, complex and interconnected, but it is clear that in certain periods drought, famine, disease and social unrest may have coincided to create regional social, economic and environmental crises in addition to or superimposed on those that were experienced nationally.

Drought, Risk and the Social Construction of Flooding in the Bajío

Although the colonial settlement of the Bajío was primarily driven by mineral exploration and exploitation, this region more than any other gained a reputation as the agrarian heartland of the colony. Efforts to reduce vulnerability to variations in annual precipitation were geared primarily toward ensuring steady and regular water supply, specifically for irrigation for cereal crops, but also for other commodities, domestic use, for livestock and also for the functioning of the mines. Attempts to maintain a supply of water, and so improve resilience to climatic variability, however, might have in fact resulted in a temporal displacement of risk. Investigating historical trajectories of vulnerability in the Bajío, for example, reveals that while water management systems effectively facilitated the development of a thriving agrarian economy in the region, they also helped to produce a society that was perhaps overly dependent on these systems. This was not so much a problem during years of normal climatic conditions, or even during single year droughts. Successive droughts, however, such as those witnessed in the 1750s and 1780s, highlighted the inefficiency of water management as a buffer against water scarcity and in fact revealed the problems of this dependency.

Archival documentation indicates that flooding in the Bajío, particularly in the urban centres of Guanajuato, Léon and Celaya, was a persistent problem. A cluster of highly destructive, well-documented and memorable floods was recorded in the second half of the eighteenth century. By this time, the level of water management, storage and diversion in the region had reached its zenith. Neglect, mismanagement and, on some occasions, deliberate sabotage of these water management structures might have exacerbated flood risk in the region. To some extent this apparent increase in flood activity in the later colonial period may also reflect other more general social and economic trends. Flooding might be a function of the expansion in mining and indeed the cumulative effect of centuries of deforestation, both contributing to soil erosion and sedimentation into the river valleys of the region. This could most certainly explain the concern over sediment build-up and the frequent references to the need to dredge and clear river channels at this time. One must also bear in mind the changing nature of risk. Population increases and economic development in the seventeenth and eighteenth centuries meant a greater number of potentially vulnerable people and an increased risk of life and economic losses associated with floods by this stage. The eighteenth century is also simply better documented than earlier periods, possibly explaining the increase in recorded floods from that period, a factor that may in fact skew any trends in the frequency of extreme events that we might elicit from the archival records in general.

Extreme or unusually heavy rainfall may have been instrumental in some of the recorded flood events. The start of a normal rainy season could have represented something of a trigger to flooding in a number of instances. Only in a couple of cases, however, was flooding directly linked by contemporary commentators to unusual or extreme rainfall or storm events. Rather, the problem of flooding in Guanajuato seems to have been perceived to be a function of the geography of the area and the imposition of water storage and management structures, though sedimentation was widely recognized as an important contributory factor. Residents who had to deal with regular inundation, and also 'experts' hired to investigate the causes of flooding, considered the majority of floods to be a function of one or more of these anthropogenic factors. As has been recognized elsewhere (Daanish, 2002: 94), such explanations of flooding may reflect a general awareness of the power relationships between different social sectors of the population and an appreciation for the differential social vulnerability to drought and flooding. More cynically, an understanding of the nature of the penalties, fines and rewards that could be secured by taking legal action might have driven some of the lawsuits between water users. As was demonstrated in chapters 4 and 5, local communities were keenly aware of their water rights and frequently contested them. By challenging neighbouring landowners' use or misuse of water sources, or charging them with damages due to dam neglect and flooding, fines could be levied and water rights modified.

Flooding prompted a range of responses by individuals and communities and, in some instances, led to quite elaborate structural projects and public works, driven or financed by local government. Although documents indicate that there may have been differences of opinion over the causes of some flood events, it is interesting to note that there were shared concerns over the effects of floods more generally. Indeed, economic and, in some situations, life losses to flood events, most particularly those that were associated with unusual weather and not attributed to human intervention, may have generated a level of social consciousness that promoted cross-cultural interaction. There is evidence that some flood experiences may have encouraged community participation in flood management and alleviation. The devastating events of the second half of the eighteenth century, for example, affected all sectors of society, irrespective of wealth, power, age, class, race or gender. The archival documentation indicates that these shared experiences and losses may have inspired a degree of collaboration and co-operation, not only in terms of immediate recovery, but also in the development and implementation of longer-term flood mitigation and alleviation schemes. Examining the various projects proposed to control or mitigate flooding in Guanajuato provides insight into the degree of interaction between different sectors of society in response to damaging and in some cases devastating events and highlights how societies were to some extent willing to tax themselves, in terms of labour and money, in order to avoid future losses. Though scholars have traditionally argued that eighteenth-century interactions between indigenous and Hispanic groups in Guanajuato were characterized by conflict and dispute over land, labour and water, these references to co-operation raise interesting questions over the way in which shared environmental values and concerns might have transcended social and class divides. This kind of co-operation gains even more significance when set against the turbulent social, economic and political context of late eighteenth and early nineteenth century Bajío in general.

Resilience and the Rare Event: Climate, Society and Human Choice in the Indigenous South

There was a good deal of indigenous land retention in Oaxaca throughout the colonial period. To some extent this was a function of the strength and power of the caciques in the region prior to European contact, but also due to the assistance they provided for the Spanish in the process of conquest. Although some lands were transferred into the hands of Spanish elites, including the lands of the Marquesado del Valle, and the various branches of the Church, the caciques remained important property owners, often renting out lands and resources to Spanish and Creole tenants, at least until their local power waned in the later eighteenth century. European commodities and livestock were of course introduced to the region, but there was relatively little actual Spanish investment in the land. Rather, the small Spanish population relied on the food and commodities supplied by tribute-paying Indio communities and remained predominantly urban in orientation. Indigenous communities were responsible for producing basic goods, while the caciques represented the main suppliers of various food products for domestic markets. The fact that the most valued commodity to be produced in the region, cochineal, was dependent upon intensive indigenous labour and local knowledge, is thought to have helped preserve the continuity of indigenous control of land. A distinctive colonial political economy thus developed in this region of New Spain in which indigenous communities and nobility figured prominently. This was to have an important defining influence on the way in which climate variability and extreme or unusual weather events were experienced.

The cochineal industry might have provided some insurance for indigenous communities. The prevalence of the cactus favoured by the cochineal insect, particularly in the southern arm of the Valley of Oaxaca, ensured an alternative income, and at times a basic livelihood for those with only limited access to good land and water supplies (Taylor, 1972: 47, 94). While cochineal became the most important commodity to be produced in late colonial Oaxaca, it also represented a vitally important product upon which indigenous communities could rely. Indeed, the arrangement of credit facilities for producers, at least until the abolition of Alcades Mayores in the later eighteenth century, allowed cochineal producers some level of financial security. As shown, however, production was sensitive to climatic variability, and the anomalous weather conditions leading up to and around the turn of the nineteenth century were considered to be among the factors which may have contributed to the decline in cochineal production around this time.

Although apparently less of a problem compared to other regions of New Spain, drought did result in harvest crises across the region on a number of occasions. There are, in addition, several incidents of more localized drought induced harvest losses throughout the colonial period.[1] Some of these agricultural losses may be associated with traditional but seemingly high risk agricultural practices in the context of the sequia intraestival between July and August. The sowing of maize and wheat in Oaxaca traditionally takes place as soon as the rainy season arrives, usually in late May or early June (Dilley, 1997). This also seems to have been the practice in the colonial period. In an item of correspondence from 17 June 1717, for instance, between the owners of the haciendas of Santa Ynes, Y and Santa Ana near to the Y River at the southern end of the Zimatlán Valley, one landlord suggests, 'with the first rainshower, I begin to sow'.[2] A June planting, however, places peak maize water use sometime in mid to late August. The occurrence of a mid-season drought could thus strike the young plants during their most moisture sensitive period and the crop could be destroyed as a result (Dilley, 1997). Some of the reported harvest losses of the colonial period might thus be explained by the normal mid-season drought striking the new seedlings. That some harvest losses may have actually been a function of agricultural practice and not solely a result of unusual weather events must, therefore, be considered.

Flooding appears to have represented more of a problem for communities and landholders in the Valleys of Oaxaca (Kirkby, 1973: 36). Not all documented floods were responses to climatic changes and extreme rainfall events. Changes in land use within river catchments might explain some of the rapid and devastating changes that took place downstream on rivers such as the Atoyac. The timing of particular rainfall events in the agrarian calendar was also influential in determining human vulnerability and hence the scale of impact that resulted. The rivers of the region, and particularly the Atoyac, appear to have frequently burst their banks when water levels were high towards the close of the rainy season. So accustomed were some riparian communities in Oaxaca to flooding at this time of the year, when water levels were high, that they would store foodstuffs in preparation. Flooding at the start of the rainy season could also result in flash flooding of some of the more ephemeral streams such as the Salado. Such events may have had greater impact as this was a time when crops had been sown but not harvested, and when there was only limited seed in storage to fall back on. Resultant flooding in this case was associated with normal climatic conditions, yet because of the rapidity and scale of impact, the event may have been reported as unusual and extreme.

There was and is considerable spatial variation in the scale and intensity of any particular event, such that the impacts of unusually heavy rainfall might not be synchronous over an area. Thus, while heavy rains contributed to flooding of the Atoyac on numerous occasions throughout the colonial period, flooding was mainly restricted to the southern reaches of the valley where the channel is shallower. The same rainfall event could, therefore, have relatively little impact further north and thus might not be considered an extreme weather event per se so much as the localized impact of heavy, but nonetheless normal, seasonal rains. There was thus ample opportunity for both over- and underestimation of the impact of rainfall events. A number of particularly dramatic floods recorded in the colonial archives, however, clearly do relate to extreme weather events. Two of these, in 1783 and 1788, are described as being associated with hurricanes. The descriptions of the devastation caused by the events of 1599 and 1721, and the fact that they are used almost as historical benchmarks in the archival documentation, lends support to the hypothesis that they too might have been manifestations of more extreme weather phenomena.

Severe frosts appear to have been relatively rare events in the Valley of Oaxaca,[3] striking mainly upland locations (Lorenzo, 1960; Taylor, 1972: 72). That it was mainly the sugar crops that were affected perhaps suggests that sugar was ill-suited to the areas where people attempted to grow it. Sugar was not a staple crop and local populations were not dependent upon its production. Selective harvest losses such as this, therefore, do not appear to have invited local government intervention or aid of any kind. Persistent failure of the sugar harvests to weather events in the valley lowlands, moreover, did not deter attempts to raise the crop. Indeed, towards the close of the eighteenth century and after Independence there was considerable expansion in the amount of land sown with sugar cane, though mainly in the southern reaches of the valleys, and in the Zimatlán valley in particular.[4] From this it could be assumed that frost was considered to have been such a rare, if admittedly potentially devastating, event that the potential advantages, should a sugar crop mature, far outweighed the perceived risks of economic loss. Prevailing socioeconomic conditions and human choice, in this case agrarian practice, therefore, seem to have been important influences in shaping the relative vulnerability of individuals or communities to climatic variability and extreme weather events in colonial Oaxaca.

Crises in Context and Historical 'Double Exposure'

Similarities have long been drawn between social and economic crisis in seventeenth-century Europe and some parts of Latin America, though rather less attention has been paid to comparable climatic records (Claxton and Hecht, 1978). As Swan (1981) suggested, however, some of the most devastating and apparently meteorological disasters of the colonial period in Mexico might be at least partially understood by turning our attention to Europe. There are, for example, a number of distinctive periods when abnormal and extreme weather conditions prevailed over Mexico and parts of Europe. As has been suggested in the previous chapters, the 1690s were particularly problematic in Mexico. There were a variety of anomalous weather events during this most calamitous decade that would result in harvest failure and dearth and may have also contributed to epidemic disease, crop pests and social unrest both in the central and northern regions of the country. Yet this period also seems to correspond to one of climatic disaster across Europe. This was a period of low solar activity now recognized as the Late Maunder Minimum, but there were various regional implications. The Alps experienced a series of cold and damp springs in

1688, 1689, 1690 and 1692, affecting harvests across a broad area. Throughout France and Switzerland, for example, 1692 was a year of extremely cold and wet conditions. On a single day in late April of that year 6 inches of snow were reported as falling in Paris, while one of the most significant famines in the history of Western Europe arrived the following year. In England too, temperatures, particularly those in winter, plummeted in the late seventeenth and early eighteenth centuries (Manley, 1974). In fact the last decade of the seventeenth century and early years of the eighteenth century in Europe were characterized by lower than normal temperatures (both summer and winter) and wetter conditions.

This anomalous weather resulted in calamitous famines in many regions. France was particularly severely affected by two years of dearth between 1692 and 1694 (Michaelowa, 2001), while Sweden too began to suffer famine by the middle of the decade (Davidson, 2004). In Scotland the effects were particularly dramatic. Grain shortages were recorded in some Scottish counties following harvest failure in August of 1695, the first harvest failure in Scotland since 1674 (Smout, 1979). The following year harvests across Scotland failed once more and by December that year the country was on the verge of famine (Davidson, 2004). Crops would fail across a broad area again in 1698 and then in some areas the following year. The traditionally mercantilist government permitted duty free food imports and grain was transported from the highlands to the lowlands. Notwithstanding these efforts, the sequence of calamities resulted in what some have referred to as Scotland's last national disaster (Wrightson, 1989: 255). Although figures are inexact and we shall never know exactly how many people died, it is estimated that 5–15 per cent of the population of Scotland, that is to say, between around 50,000 and 150,000 people, may have perished in this disaster, while there was a noticeable increase in clan feuds, robbery and looting at this time and considerable population displacement in search of food (Davidson, 2004). The unusual weather of the 1690s revealed the extreme vulnerability of Scotland's agricultural sector, still dominated as it was by a feudal tenurial system. This had prevented any advances in commercialization and increased productivity and placed unreasonable rental demands on impoverished tenants (Davidson, 2004). Here was a 'terrible instance of the vulnerability of a primitive economy to bad weather' (Smout, 1979: 225).

There has also been extensive coverage of the degree of climatic variability across Europe in the second half of the eighteenth century. In Mediterranean Europe, the period between 1760 and 1800 saw an increased frequency of both drought and flood episodes thought to be associated with the so-called Maldá climatic anomaly (Barriendos and Llasat, 2003). The 1780s in particular are thought to have been a period of 'very pronounced regional anomalies in the general circulation' across Europe generally (Kington, 1988: 2).

An apparent increase in the frequency of cyclonic and anti-cyclonic days contributed to harvest failures across France (Kington, 1980) and as Cobb (1970) noted, the consequent problems of food scarcity might have represented key influences on popular behaviour, contributing to the level of dissent that would culminate in revolution later in the decade. Whether the climatic conditions that influenced weather phenomena in France at this time were also responsible for the unusual phenomena recorded across Mexico in the mid-1780s is a hypothesis that remains to be tested.

Some links have been highlighted between an increase in droughts in Mexico and the cold, wet conditions that prevailed over Europe during the period of the Little Ice Age (Swan, 1981; Gibson, 1982; Hodell et al., 2005b). This has led to a tentative suggestion that a 'Little Drought Age' might in fact have prevailed over Mexico, and perhaps more broadly across lower latitudes, during this period (Endfield and O'Hara, 1997). Though many studies of climate change in Mexico have been conducted in recent years, the paucity of available historical Mexican climate data relative to that for Europe with which to test this association means that such links still remain hypothetical.

It is to be expected that at least some of the individual drought, flood and storm events recorded in the colonial archives of Mexico might also be explained by historical ENSO activity (Mendoza et al., 2005). Ortlieb (1999), for instance, has identified medium El Niño events in both 1784 and 1785, possibly providing one potential explanation of the droughts and frosts recorded across Mexico at this time. The droughts and harvest failures recorded across the Bajío in 1641 and those recorded in north and central Mexico between 1750 and 1751 also correspond to strong or moderate El Niño years according to Quinn and Neal's (1992) El Niño chronology. The reliability of the documentary data upon which such classifications were based, however, has been questioned (Hocquenghem and Ortlieb, 1992; Ortlieb, 1999). Moreover, one needs to consider that there would most probably have been a time lag between the weather event and the emergence, recognition and recording of a crisis. Other recorded events, however, appear to have been synchronous in different parts of the country and to have coincided with more widely documented anomalous weather. Unusually heavy rain and flooding recorded in Guanajuato and elsewhere in the country between 1791 and 1792, and also the poor harvests reported in Oaxaca the subsequent year,[5] might well be associated with the well-documented El Niño of this period. Elsewhere, as far afield as India, St. Helena and Montserrat, there were simultaneous droughts in 1791 and the period 1789–93 is particularly well documented as being one of global climatic abnormality generally (Grove, 1998).

Quinn and Neal's record suggests there was a strong El Niño in 1783. Between 1783 and 1784, freak weather was reported nationally in Mexico and there was growing recognition of changing weather conditions and patterns and possible links with deteriorating human health across the country. A whole series of unusual phenomena was reported in Oaxaca in this year. It is perhaps interesting to note also that the Lakagigar eruption (Iceland), which took place between June 1783 and February 1784, resulted in anomalous weather events across north and west Europe and North America (Grattan and Brayshay, 1995; Stothers, 1996; Witham and Oppenheimer, 2005). Reports from the time indicate the occurrence of 'dry fogs' or sulphurous haze, unusually warm summers in Europe and North America and a cold winter in 1783 and unusually cold temperatures over the succeeding two to three years, characteristics which Stothers (1996) argues were due to the effect of aerosols from the eruption. It is also possible that 'dry fogs' might have also been experienced in more easterly regions. In 1783 Chinese chroniclers of Henan province, for example, reported severe fog and dark daytime skies (Demarée et al., 1996). The volcanic eruption in Asama, Japan, also in 1783, may have contributed to these atmospheric impacts. But one could perhaps reconsider the anomalous weather conditions across different regions of Mexico in 1783, 1784 and 1785 in light of the widespread impacts of the volcanic activity of this year.

It is clear that a number of periods of unusual and anomalous weather affected Mexico and might have contributed to periods of social and economic crisis, and particularly, it seems, in the eighteenth century. Some of the events and crises documented in the colonial archives at this time, those of the 1780s in particular, might relate to global climate phenomena. However, caution is obviously needed in drawing links between the conditions described in Mexico and events further afield, especially given the high degree of normal climatic variability experienced in Mexico generally. Furthermore, and perhaps most importantly, the crises that affected Mexico at this time were not solely a result of meteorology. The different regions of the country were inevitably confronted by unusual and in some cases extreme weather which resulted in various impacts, but a range of non-climatic parameters combined to render society as a whole potentially more vulnerable to these changes and hence also to subsistence crisis in the eighteenth century. Population expansion coupled with increased economic development effectively meant that there was more disaster potential, while the very hierarchical structure of Mexican society meant that some groups of people were disproportionately affected during periods of crisis or following a disaster.

One could also apply O'Brien and Leichenko's (2000) terminology to eighteenth-century Mexico and consider the implications of 'double exposure', that is to say, the coeval impacts of climate change and the implications of a period of globalization. As Ponzio (2005: 437) has suggested, and as discussed in the previous chapter, the eighteenth century was 'characterized by increasing market integration and a growth in international trade'. The Bourbon Reforms and their orientation towards free trade certainly stimulated commercial growth and sought to challenge the power of local elites and reassert Crown control. A focus on external trade undermined local producers, while the removal of the Alcades Mayores resulted in the withdrawal of credit facilities for many indigenous communities. This economic growth masked a number of structural problems. There was an increasingly dependent, indebted underclass, at least in the agrarian heartland of the Bajío, that lacked access to a means of securing its own food supply. In addition, demographic expansion through much of the century placed pressure on limited lands and resources, and stimulated more rural-urban migration. The lack of urban employment opportunities may have contributed to an increase in crime and unrest. All sectors of the population were affected by a number of subsistence crises during the 1700s, and particularly those of the second half of the century. The synergy and simultaneity of periods of unusual and extreme weather and globalization processes, however, might have contributed to this outcome, and as such might have rendered society 'doubly exposed' to climate variability by the eighteenth century. For this reason, this period may appear to be more calamitous when viewed through the medium of the historical archives.

Closing Comments

If experience with weather and climate in the historical and recent past may go some way to aiding our understanding of the implications of climate change in the future, then there is a need to develop a more comprehensive assessment of changes in weather and extreme climate events over longer time scales. Indeed, current concern over vulnerability to global warming and potential increases in unusual and extreme weather events, including those associated with the El Niño Southern Oscillation, has highlighted the need for long-term but temporally detailed climatic histories. It is becoming increasingly clear, however, that explorations of the impacts of and social response to these changes are equally if not more important. Moreover, it is essential to try to obtain a better understanding of how societies have conceptualized these changes and endeavoured to make themselves effectively more resilient to them (Hassan, 2000). As Butzer (2005) has illustrated, human perception and, in particular, ecological behaviour are absolutely pivotal to understanding cause and effect in environmental or climate change scenarios. In this context, investigations of social vulnerability to climate variability in historical perspective should be regarded as being central to future climate change studies. Historical archives represent one medium through which such investigations might proceed.

Documents such as those used in this study are rarely free from subjectivity and the derivation of climatic or cultural information from them is fraught with difficulty and marred by error. Periods of crisis, by definition, are more noteworthy than normal conditions. For this reason, historical records tend to be biased towards describing and depicting the impacts of disasters and crises perceived to be of greatest magnitude or which are thought to have resulted in greatest economic disruption and/or life losses. Archival records may thus serve to illustrate a society disproportionately sensitive and more poorly adapted at particular times than it actually is. Routine and long-term socio-cultural adaptations in contrast, perhaps the most meaningful information with which to investigate the relative resilience of a particular group of people in a given area, tends to be 'hidden' (Meyer et al., 1998). For this reason, the interpretations presented in this study necessarily remain partial at best.

This study has also focused on three case study regions within Mexico. It is recognized that there was and still is considerable variation in socioeconomic and environmental characteristics within as well as between each of these regions and between them and other regions of the country. In this sense, it might be argued that few generalizations and interpretations can or should be drawn from archival investigations of historical changes in these case study regions alone. Nevertheless, it is hoped that the rich colonial archives of Mexico have been highlighted as a vital resource with the potential to provide insight into the way in which societies understood vulnerability to climatic variability and attempted to make themselves more resilient to its implications.

Notes

1   In October 1999, for example, a tropical depression brought extraordinarily heavy rain. Homes, cattle and crops were swept away by flash floods (*New York Times*, 15 October 1999), while mudslides were reported across central and southeastern Mexico (*Miami Herald*, 25 October 1999). An estimated 360 people lost their lives in Puebla, Veracruz, Hidalgo and Tabasco, though in total nine states across the country were affected. The events unfolded at a time when communities were still recovering from an earthquake on 30 September that year and which took its toll particularly on communities in Oaxaca. About 200 municipalities in the states of Puebla, Veracruz and Hidlago were affected by flooding and mass movement processes that resulted from a tropical depression from the Atlantic Ocean in December 1999. In October the preceding year, Hurricane Mitch had killed an estimated 10,000 people and had caused billions of dollars in damage. Hundreds of slope failures were triggered by intense precipitation, which in some locations reached 420 mm during a 24-hour period, resulting in 263 deaths (Alcantara-Ayala, 2004).

2   Saunders (2000) considers the origins of the term 'environmental refugees' as a political and environmental construct.

3   See, for example, Camuffo and Enzi (1992); Grove (1983); Barriendos (1997); Brazdil et al. (2003, 2005); Barriendos and Llasat (2003); Pfister (2005); Pfister and Brazdil (1999); Pfister et al. (1999, 2001); Martinvide and Vallve (1995); Piervitali and Colacino (2001); Nordli (2001); Retsö (2002); Rodrigo et al. (1998); Pfister (2005).

4   See, for example, Metcalfe (1987); O'Hara and Metcalfe (1995); Endfield and O'Hara (1997); Gioda and L'Hôte (2002); Prieto (1983); García-Herrera et al. (2003); Prieto et al. (2000, 2004); Chenoweth (1998, 2005).

5   Nicholson (1979); Lindesay and Vogel (1990); Vogel (1989); Nash and Endfield (2002, 2002b).

6   See, for example, Zhang (1994); Liu et al. (2001).

7   Garcia- Herrera et al. (2001); Garcia-Herrera et al. (2005).

8   These sources have long been used to investigate the demographic changes of the pre- and post-Conquest period (Cook and Borah, 1960; Williams, 1972), adjustments in land tenure (Prem, 1984, 1992; Licate, 1981; Pellicer, 1994) and agricultural dynamics (Simpson, 1952; Taylor, 1972; Sanders, 1992) since the Conquest.

9   The ramos referred to in tables 1.1 and 1.2 represent the main sources used in this study, though others were also consulted. These will be referenced as and when they are cited.

10   Particular publications consulted included Domingo Lázaro de Arregui (1946), Nuñez Cabeca de Vaca (1958), Mota y Escobar (1940), Tamaron y Romeral (1937), Lafora (1939) and Neumann (1969).

11   Documents were consulted in the following archives: Archivo General de la Nación, Mexico City (AGN); Archivo General del Estado de Oaxaca (AGEO); Archivo Historico Municipal de la Ciudad de Oaxaca (AHMCO); Historico Municipal de Leon, Guanajuato (AHML); Archivo Casa de Morelos, Morelia, Michoacán (ACM); Archivo Historico Municipal de Fondo Colonial, Chihuahua (AHMCH); Archivo Hidalgo de Parral, Chihuahua (AHP); Archivo de las Indias (Seville) (AGI); Archives held in the Pátzcuaro (Michoacán) archives (P). Archives citations in the remainder of the text are referenced in the following manner: the abbreviation of the archival repository, the document group (ramo) consulted, the volume number (Vol.), the expediente (Exp.) or legajo (Leg) number (if applicable), and/or the page (foja) number, denoted by Fa (single pages) and Fs (multiple pages). (Page numbers may also be accompanied by f (frente) facing page or v (verso) reverse page). References for documents held in AHP are referenced according to microfilm (upper case letter) and item (lower case letter).

CHAPTER 2   CLIMATE, CULTURE AND CONQUEST: NORTH, SOUTH AND CENTRAL MEXICO IN THE
            PRE-EUROPEAN AND CONTACT PERIOD

1   The early, middle and late Holocene periods are thought to have been punctuated by periods wet enough to establish large pluvial lakes in currently dry basins in the Chihuahuan Desert. This increased precipitation, cooler temperatures and reduced evaporation might be associated with southward shifts in winter storm tracks, which are related to long-term ENSO variability during the Holocene (Castiglia and Fawcett, 2006).

2   This period was followed by a wetter than normal decade between 1477 and 1486 (Cleaveland et al., 2003).

3   Chihuahua was actually initially known to the Spanish through Nuñez Cabeza de Vaca, one of the first Europeans to travel through the area, albeit while escaping from six years as a hostage in Texas. Cabeza de Vaca provided one of the first European accounts of the region.

4   Investigations at two sites in Guanajuato – Carabino and El Corporal – have revealed evidence of ceremonial platforms and ball courts which bear similarities

to those at Tula. The Toltec cultural tradition is also evident in the Valley of Querétaro, in El Pueblito, where new buildings were constructed on top of previous foundations.

5   The Chichimecas' or dog indians' frequent attacks on early settlements presented one of the main obstacles to Spanish colonization of the area.

6   Gonzalo de las Casas 'Noticia de los chichimecas y justicia de la guerra que se laes ha hecho por los españoles'. In Herman Trimborn (ed.) (1936) *Quellen zur Kulturgeschichte des prakolumbischen Amerika* (Stuttgart): 154 (cited in Powell, 1945: 324).

7   The Valley of Oaxaca actually features as a significant supplier of foodstuffs in Aztec tribute lists (see Blanton et al., 1982).

8   The investigation involved the transcription, translation and interpretation of two primordial titles, written in Mixtec and Nahuatl languages and a large map. Primordial titles are among the most ambiguous forms of indigenous writing in the local and national Mexican archives, often being presented in response to Spanish demands for proof of title to land. Many are thought to have been falsified or forged and few in fact pre-date the middle of the seventeenth century. The titles in question here were produced in relation to a dispute over territory between two indigenous communities in the Valley of Oaxaca and ostensibly written in 1520. They were presented as evidence of primordial title to the Spanish authorities in the 1690s (Sousa and Terraciano, 2003).

9   The Zapotecs were the primary occupants of the region, but from about the year 1350 they had been under assault by Mixtecs moving in from the east. Oaxaca had been conquered by the Aztecs from Tenochtitlán (i.e. forced to acknowledge its primacy and send tribute) shortly before 1500 (Taylor, 1972: 22).

10 AGN Hospital de Jesus, Leg 389, Exp. 5.

11 Relaciones Geográficas de Oaxaca en Paso y Troncoso Franc (1939–40) *Papeles de la Nueva España*, 2nd Series Vol. 4: 288–300; Miahuatlán, 1609.

CHAPTER 3   EXPLORING THE ANATOMY OF VULNERABILITY IN COLONIAL MEXICO

1   After about 1786, however, and during, in fact, one of the most devastating subsistence crises in Mexican colonial history, each male member of the Republica de Indios aged between 18 and 50, irrespective of whether they were a head of household or not, became eligible to pay tribute every year (Ouweneel, 1991: 541).

2   Each caballería of land measured 42.8 hectares/265 acres

3   Repartimiento, in this instance, refers to the colonial labour system imposed upon the indigenous population of Spanish America. A conquistador would take over and supervise a number of indigenous workers, who would provide their labour in the agricultural production or in the mines.

4   Cortés had witnessed the dramatic demise of the indigenous populations of the Antilles as a result of disease and overwork in the mines. Although he had exempted tribute Indios from working in the mines themselves – an occupation reserved for slaves and prisoners of war – they were obliged to provide food for the miners as well as tribute commodities for their Spanish overlords.

5   The new reforms were adopted on a broader scale after 1573.
6   AGN Mercedes, Vol. 7, Fa 125.
7   AGN Mercedes, Vol. 9, Fa 113v.
8   AGN Mercedes, Vol. 11, Exp. 6.
9   AGN Mercedes, Vol. 8, Fa 15v (cited in Butzer and Butzer, 1993).
10   AGN Mercedes, Vol. 1, Fa 233 (1542); Vol. 3, Fa 135 (1550).
11   The 'sin perjucio de tercero' clause included in most merced awards and referred to above was expanded to specifically refer to 'Indios or other third parties', according to the revised *Recopiliación de Leyes de las Reynas de Indias* – set of guidelines or constitution – of 1594. (*Recopiliación de Leyes de las Reynas de Indias* (1681), Conséjo de la Hispanidad, Madrid, 1987.)
12   The allotment was increased to 600 varas (504 m) according to a decree dated 1687. The legal fund to which indigenous communities were entitled was measured from the last house in the town and 'in all four wind directions', north, east, west and south. This total amount of land is thought to have been in the region of between 80 and 100 hectares, depending on whether the measurements were made in a circle or a square around the settlement. The Real Cédula of 1695 modified the location from where these stipulated measurements were made. Instead of starting from the last house in the town, the land measures now began from the church, normally located in the centre of the town (Florescano, 1972). See also Fabila (1941).
13   *Recopiliación de Leyes de las Reynas de Indias* (1681): Vol. 2, Libro VI, titulo. 3, ley VIII, Fa 209.
14   The *Recopiliación de Leyes de las Reynas de Indias* provided a number of provisions for water rights, based to some extent on the *Siete Partidas*, the thirteenth-century legal code which informed much of Spanish legislation. This Castillian legislation was transferred to New Spain.
15   Over the past five decades there have evolved two schools of demographic thought in this respect: one group of scholars, collectively referred to as the Berkeley School, tended to support theories of very large pre-Columbian populations in the Americas which were virtually wiped out by European pathogens (Borah and Cook, 1963; Cook, 1949; Cook and Borah, 1957, 1960, 1968, 1971, 1974, 1979; Cook and Simpson, 1948). Others, more sceptical of the high population estimates, doubt the data sources and their interpretation (MacArthur, 1970; Sanders, 1992; Rosenblat, 1992), favouring lower estimates and less catastrophic population decline.
16   Attempts have been made to identify the different diseases involved in specific recognized epidemics, based on contemporary European and indigenous accounts (Joralemon, 1982: 115; Prem, 1992: 23). Indigenous nomenclature of diseases has also been used to draw comparisons between individual phases of disease and to trace their reappearance (Prem, 1992; Acuña-Soto et al., 2000). Descriptions rarely drew on any formal medical training, so these testimonies need to be interpreted with a degree of caution. Moreover, to these difficulties must be added a degree of uncertainty over the very many and often conflicting diagnoses of specific disease outbreaks (Prem, 1992: 23).

17 Discussed in Solórzano y Pereyra (1629–39), *Politica Indiana* Lib. V, Cap. XII; Pinelo, L.P. *Tratado de confirmaciones reales de encomiendas, oficios y casos en que se requieren para las Indias Occidentales*, Madrid.

18 The legal fund to which the Indio populations were entitled was considered by the communities themselves to be insufficient. Many of the lawsuits over land filed in the eighteenth century, for example, highlight that pueblos and Indio communities were no longer in possession of this legal allotment as a result of hacienda encroachment.

19 AGI Guadalajara 55, no. 8 (1574).

20 Such was the nature of the labour shortages that Francisco de Ibarra, first governor of Nueva Vizcaya, made a request of the Crown to send black slaves to the colony to work in the mines (AGI Patronato 20, no. 16, letter from Francisco de Ibarra, 1576).

21 AHP 1637B; 1644A; 1625B; 1654Ab; 1658Aa; 1699b; see Griffen (1979: 47).

22 AGI Guadalajara 65, no. 60: Relación de Frey Rangel (1578), cited in Cramaussel (1990a: 38).

23 AHP Chihuahua 1685 Aa.

24 AGN Misiones, Vol. 27, Exp. 6 Fs 455–483.

25 AGN Misiones, Vol. 27, Exp. 5 Fs 392–395.

26 AGN Misiones, Vol. 27, Exp. 6 Fs 455–483.

27 AGN General de Parte, Vol. 37, Exp. 30.

28 See, for example, AGI Mexico, Carta del Virrey al Rey, 19 October 1577, cited in Gerhard (1982: 242); Report of Jesuit visitor, Andrew Perez de Ribas, June 13, 1617, and report of P. Diego de Alejos, Tecguciapa, 18 May 1617, in AGN Archivo Provisional Misiones, Caja 2; Report of P. Diego Ximénez, San Andres, AGN Jesuitas III-15.

29 As the number of Spaniards living in the area increased, encomenderos often rented out their labour. Indeed, repartimiento drafts actually increased in this region at a time when its abolition was being considered by the viceroyalty (Deeds, 1989). Not until 1670 was a royal decree issued to the governor of Nueva Vizcaya which forbade the practice of repartimiento under the encomienda system.

30 AGI Audiencia de Mexico, Leg. 20 (letter from the Viceroy to the king, 19 October, 1577).

31 AGI Guadalajara 35 (1579).

32 Based on a description found in Mota y Escobar, A (1940 edition).

33 Fanega: grain measure. One fanega is approximately 1.6 bushels.

34 AGI Guadalajara 28, censo de 1604 in Borah (1955).

35 AHP 1654a, 115–134, in Cramaussel (1990b: 136).

36 Deeds highlights various disputes involving the Cortés del Rey Mayorazgo as recorded in the Tierras of the Archivo de Hidalgo de Parral and the Archivo General de las Indias, Audiencia de Guadalajara, Leg. 120 (Deeds, 1989: 438).

37 Documentos para la Historica de Mexico, 1668, 4-III, 223–230, 259 (Deeds, 1989).

38   A number of such 'donations' are detailed in Memorial del P. Procurador
       General Bernardo Rolandegui, ca. 1704, AGN, Archivo Histórico de Hacienda,
       Temporalidades Leg. 325, Exp. 51, cited in Deeds (2003: 20).
39   AGN Temporalidades, Vol. 196, Fa 306v.
40   One in Marfil, one at Tepetapa, a third at Santa Ana and finally one at El
       Puktimi, situated in the fold of the hill of Cuarto (Marmolejo, 1967: 118).
41   AGN Mercedes, Vol. 4, Fa 532, cited in Powell (1944).
42   AGN Mercedes, Vol. 4, Fa 186.
43   Powell (1944: 183) notes that the Viceroy himself had originally intended to
       supervise the founding of the town, but he was prevented from doing so by
       illness while he was travelling north.
44   AGN Media Anata, Vol. 35, Fs 244–249.
45   AGN Mercedes I VII: 573–574, cited in Powell (1944).
46   AGN Media Anata, Vol. 35, Fs 244–249, verso.
47   AGN Mercedes V: 87, cited in Powell (1944).
48   According to Ciudad Real, A (1976 edition) *Tratado curiosos y docto de la grandez
       de la Nueva España. Relación breve y verdadera de algunas cosas de las muchas que
       suciedieron al padre Fray Alonso Ponce en las provincias de la Nueva España siendo
       comisario general de aquellas partes.* UNAM, Institute Investigaciones Historica.
49   AGN General de Parte II, fol. 63v, cited in Powell (1944: 198).
50   AGN Mercedes, Vol. 1, Exp. 53.
51   AGN Mercedes, Vol. 5, 120v; AGN Tierras, Vol. 675, Exp. 1 Fa 75v; Vol. 136,
       parte 1a; see also, Martinez de la Rosa (1965).
52   Between 1542 and 1550, for example, Bocanegra was conceded 18 additional
       mercedes, each of a cabellería and a half for himself and his sons (Wright Carr,
       1989).
53   AGN Tierras, Vol. 741, Exp. 1 Fa 55.
54   On 9 April 1581, for example, Viceroy Mendoza granted a merced for an estan-
       cia for cattle and two caballerías of land, to a Francisco de Salcido. Less than a
       month later, on 2 May, Salcido passed the merced on to Antonio Aldrete, a
       resident of Mexico City, while on 2 January the following year, Aldrete sold the
       property on to Francisco de Vallejo, also from the capital (AGN Tierras, Vol. 50,
       Exp. 1 Fa 14).
55   AGN Tierras, Vol. 95, Exp. 6.
56   AGN Tierras, 674, Fa 45.
57   AGN General de Parte I, fol. 244, cited in Powell (1944: 191).
58   Relaciones Geográficas de la Diócesis de Michoacán, 1579–1580, 2 vols, in
       Papeles de la Nueva España, Guadalajara, Colección, Siglo XVI, 1958: 52–55.
59   AGN General de Parte, Vol. 1, Exp. 1304.
60   AGN General de Parte, Vol. 2, Exp. 60; Vol. 4, Exp. 266; Vol. 6, Exp. 794.
61   AGN General de Parte, Vol. 5, Exp. 226.
62   AHML, Leg 11, Fa 1.
63   AGN Tierras, Vol. 988, Exp. 2, Fa 92.
64   AGN Civil, Vol. 286.
65   AGN Merecdes, Vl. 8, ff 2, 29, 40v, 43, 55, 87, 100v, 111, 113, 118, 190.

66   See, for example, numerous references to 'waterlifts' or norias in AGN Historia, Vol. 72, Exp. 9; AGN Civil, Vol. 73, Exp. 3; AGN Tierras, Vol. 514, Exp. 1, Cuaderno. 2, Fa 47; AGN Tierras, Vol. 618, Exp. 1, Cuad. 3, Fa 61; AGN Tierras, Vol. 1353, Exp. 1, f. 69.

67   AGN Tierras, Vol. 2705, Exp. 3, Fa.1; AGN Mercedes, Vol. 10, Fa 3.

68   Urquiola, not dated: 7.

69   Boletín AGN, Vol. VI: 5.

70   The channel was finally constructed to the north of the Rio Lerma. To the south, the canal referred to as Brazo Moreno had already been completed for the same purposes sometime in the first decade of the seventeenth century.

71   AGN Tierras, Vol. 674, Cuad. 2, Fa 1139v.

72   AGN Tierras, Vol. 353, Exp. 2 Fa 25.

73   AGN Indios, Vol. 33, Exp. 338; Vol. 5, Exp. 128.

74   AGN Indios, Vol. 5, Exp. 598.

75   Sometimes also referred to as sobras or demasias and meaning the waters remaining after an upstream landowner had irrigated his land.

76   AGN Tierras, Vol. 2680, Exp. 29.

77   AGN Tierras, Vol. 2680, Exp. 29.

78   AGN Caminos y Calzadas, Vol. 2, Exp. 8 Fs 115–122.

79   15.9 per cent were classed as artisans, 14.9 per cent were employed in manufacturing, 9.8 per cent worked in the mines, 8.5 per cent were classed as merchants and traders, 0.9 per cent were employed in administrative occupations and 0.8 per cent were classed as professionals, military employees or members of religious orders (Gomez and Armando, 1995: 308–314).

80   This privilege was disputed by the Crown as residing solely with it and brought to a halt in 1555 (AGN Mercedes, Vol. 4, Exp. 141).

81   AGN Indios, Vol. 5, Exp. 698.

82   AGN Mercedes, Vol. 131; Vol. 13, Exp. Fa 179; Fa 178; Fa 101; Fa 115; Vol. 17, Fa 87. To some extent this delay in land investment for agriculture was due to a monopolization of the lands of the area by Hernan Cortez, whose original title to the Marqués del Valle de Oaxaca afforded him jurisdiction over the majority of the valley (AGN Hospital de Jesus, 293. Exp.133 Fa 1531).

83   AGN Mercedes, Vol. 2, Exp. 260.

84   AGN Hospital de Jesus, Leg 404 Exp. 2; Hospital de Jesus 432 Exp. 5.

85   AGN Hospital de Jesus, Leg 404 Exp. 2; AGN Tierras, Vol. 413, Exp. 2 Fa 79; Vol. 791, Exp. 2 Fa 150; Vol. 1583 48, Exp. 6 Fa 244.

86   AGN Mercedes, Vol. 24, Fa 182; Vol. 27, Fa 54; Vol. 16, Fa 121; Vol. 2, Fa 264; Vol. 53, Fa 165 y 167; Vol. 64, Fs 63 and 64; Vol. 17, Fa 147; Vol. 58, Fs 74 and 114; Vol. 18, Fa 43; Vol. 26, Fa.13; Vol. 26, Fa.13; Vol. 72, Fs 167 and 171; Vol. 41, Fa 45–46; Vol. 16, Exp. 436, Fa 119; Vol. 23, Fa 195; Vol. 40, Fa 128; Vol. 66, Fs 93 y 94; Vol. 17, Fa 87; Vol. 21, Fa 157; Vol. 41, Fa 18; Vol. 21, Fa 160; Vol. 8, Fa 92; Vol. 53, Fa 160; Vol. 3, Exp. 670; Vol. 18, Fa 146; Vol. 18, Fa 75; Vol. 18, Fa 204.

87   AGN Mercedes, Vol. 20, Fa 47; Vol. 18, Fa 124.

88   Zarate, cited in Paso y Troncosco (1939–40) *Epistolario de Nuevo España 1505–1818*, Vol. 4. 141.

89   Relaciones Geograificas de Oaxaca (various entries) in Paso y Troncosco (1939–40).
90   AGN Tierras, Vol. 2764, Exp. 33 Fa 396, No. 2088 Cat 2088 (Map); AGN Tierras, Vol. 2764, Exp. 33 Fa 396; AGN Mercedes Vol. 35, Fa 81; Vol. 34, Fa 146; Vol. 65, Fa 18; Vol. 56, Fs 12–24; AGN Vol. 72, Fa 194; Vol. 13, Fa 206; Vol. 72, Fa 215; Vol. 47, Fa 227; Vol. 80, Fa 129; Vol. 62, Fa 108; Vol. 55, Fa 19; Vol. 53, Fs 106 and 110; Vol. 64, Fa 188; Vol. 72, Fa 200; Vol. 64, Fs 197; Vol. 81, Fs 14 and 14v; Vol. 55, Fa 5; Vol. 37, Fa 84; Vol. 53, Fa 108; Vol. 37, Fa 29; Vol. 40, Fs 113–117; Vol. 67, Fa 86; Vol. 81, Fa 153 y Fa 166; Vol. 81, Fa 155; Vol. 64, Fs.10–11; Vol. 80, Fa 34; Vol. 22, Fa 230; Vol. 70, Fa 115; Vol. 81, Fa 163; Vol. 64, Fa 105; Vol. 73, Fs 74 a 75; Vol. 79, Fa 17; Vol. 31, Fa 265.
91   AGN Historia, Vol. 31, no. 3.
92   AGN Historia, Vol. 31, no. 3.
93   Relaciones Geográficas de Oaxaca in Paso y Troncoso Franc (1939–40). Papeles de la Nueva España, 2nd Series, Vol. 4; Cuahuitlan, 1580, 155–161; Chichicapa, 1580, 116–119; Cuahuitlan, 1580; see also AGN Tierras, Vol. 450, Exp. 1; AGN Salinas, Vol. 15, Exp. 14.
94   Relaciones Geográficas de Oaxaca, en Paso y Troncoso (1939–40). *Papeles de la Nueva España*, 2nd Series, Vol. 4: 301–307.
95   AGEO Real Intendencia I, Leg. 4, Exp. 43.
96   AGN Industria y Comercio, Vol. 20, Exp. 6 Fa 237.
97   In this context, the term repartimiento refers to repartimiento de bienes, the purchase and sale of goods by Spanish magistrates to their Indio charges. This institution, which is essentially a trading relationship between Indio and Spaniard, is completely unrelated to the Indio labour draft, repartimiento de Indios and referred to earlier in this chapter (Baskes, 2005: 186–187).
98   Cochineal constituted between 4 and 38 per cent of Mexico's total exports.
99   AGN Historia, Vol. 31, no. 3
100   1 arroba = 25 lbs
101   AGN Industria y Comercio, Vol. 20, Exp. 6 Fa 237; AGN Historia, Vol. 31, no. 3.
102   Between 1750 and 1821, repartimiento loans were provided at a stable rate of 12 reales (1.5 pesos) per pound of cochineal harvested.
103   Baskes (2005: 193) refers to the work of Wright (1986) in his study *Old south, new south: revolutions in the southern economy since the civil war* (Basic Books, New York), which emphasizes the dependence that southern sharecroppers might have had on credit facilities to ensure security throughout the year.
104   Producers could also escape repayment and delivery by simply leaving town when creditors arrived to claim the product. (Baskes, 2005; AGI Audiencia de Mexico, Leg. 2588, 1784).
105   Mexicans born of Spanish descent.
106   AGN Hospital de Jesus, Vol. 380, Exp. 9 Fa 85r.
107   AGI Audiencia de Mexico, Vol. 357.
108   The Bishopric at this stage corresponded to all of the modern-day state apart from Huahuapan, Xiuxtahuacan and Tzilacayoapan, which fell into the jurisdictions of Puebla, Veracruz, Guererro and Tabasco; Paso y Troncoso (1966),

La division territorial de la Nueva España en 1636. *Lecturas Historicas Mexicanas* 2: 582; cited in Taylor (1972): 174.

109   AGI Audiencia de Mexico, Vol. 881.

110   There were of course some exceptions. The monastery at Etla accumulated mostly arable land, although by the end of the eighteenth century most of this land was also rented out (Taylor, 1972: 180).

111   *Primer censo de población de la Nueva España, 1790*: 141–147. In Baskes (2000: 14).

112   Congregación was a process of consolidation of indigenous towns. Implemented following the waves of epidemics and indigenous depopulation, it was designed to collectivive scattered residual indigenous communities for ease of political and administrative control.

113   See, for example, AGN Tierras, Vol. 2680, Exp. 24 Fs 21; Vol. 165, Exp. 2 Fs 262; Vol. 182, Exp. 1 Fs 380; Vol. 1864, Exp. 2; Vol. 95, Exp. 4 Fs 5; Vol. 236, Exp. 6; Vol. 2947, Exp. 99 Fs 4; Vol. 384, Exp. 3 Fs 24; Vol. 817, Exp. 9 Fs 40; Vol. 2931, Exp. 1.

114   AGN Tierras, Vol. 1082, Exp. 1; AGN Tierras, Vol. 1268, Exp. 1; AGN Tierras, Vol. 1433, Exp. 1; Vol. 1441, Exp. 24; Vol. 1426, Exp. 3.

115   Although rental agreements were still being drawn up by caciques until the turn of the nineteenth century (Taylor, 1972: 208), reflecting their level of control, their power in the region is thought to have begun to wane in the late eighteenth century.

116   See, for example, AGN Tierras, Vol. 187, Exp. 2 Fa 327; Vol. 1110, Exp. 18; AGN Indios, Vol. 5, Exp. 154; AGN Indios, Vol. 5, Exp. 598; AGN Correspondencia de Virreyes, Vol. 12 (series 2), Fa 244; AGN Tierras, Vol. 674, Exp. 1 Fa 30; follow the dispute in Vol. 675, Exp. 1; AGN Tierras, Vol. 192, Exp. 1; Vol. 586, Exp. 8; Vol. 1872, Exp. 15 Fa 10; Vol. 671, Exp. 3, 18 ff; Vol. 2901, Exp. 36; Vol. 2959, Exp. 141; Vol. 988, Exp. 1, 2 y 3, Fa 516; Vol. 1110, Exp. 18; Vol. 1166, Exp. 1 Fa 450; Vol. 2963, Exp.116 f 246–308; Vol. 1352, Exp. 1; Vol. 1368. See also Endfield et al. (2004a).

117   AHMCO Actas de Sesiones del Cabildo 1728–1733; AHMCO Actas de Sesiones del Cabildo 1746–1748; AGEO Real Intendencia, Leg 1 Exp. 46 4 ff; AGN Industria y Comercio, Vol. 20, Exp. 6 fa 237; AGN Hospital de Jesus, Leg 118, Exp. 13; See also Endfield et al. (2004b).

118   AHMCH Hacienda, Caja 7, Exp. 4; AHMCH Jesuitas, Leg II-9, Exp. 12; AHMCH Gobierno, Caja 20, Exp. 5; AGN Mercedes, Vol. 76, Fs 95–100; AHMCH Guerra, Caja 1, Exp. 8; Caja 1, Exp. 7; AHMCH Gobierno, Caja 30, Exp. 23; AHMCH Notarias, Abasto de Carne, Caja 44, Exp. 20; AGN Ayuntamientos 173, cuaderno 3; AGN Presidios y Carceles, Vol. 4, various fojas; AHMCH Caja 40, Exp. 5; AGN Historico de Hacienda, Vol. 296, Exp. 1; AHMCH Justicia, Abasto de Carne, Caja 126, Exp. 10; AHMCH Hacienda, Caja 55, Exp. 12; AHMCH Gobierno, Caja 49, Exp. 13; see also Endfield and Fernández-Tejedo (2006).

119   AGN Tierras, Vol. 1861, Exp. 7; AGN Salinas, Vol. 15, Exp. 14.

120   AGN Obras Publicás 17, Exp. 10 Fa 43; Descripción de la Ciudad y Real de Minas de Guanajuato por José Hernandez Chico, 1788, Archivo de la Marina, Museo Naval de Madrid, Ms 563 in *Descripciones economicas regionales de la*

    Nueva España 1766–1827, Florescano, E. and Gil, I. (eds) Mexico, SE-INAH, 1975 (Fuentes para la historia economica lll) 30.
121 AGN Alcade Mayores, Vol. 1, Exp. 309 Fs 439–441; AGN Tierras, Vol. 2071, Exp. 1 Fs 1–110; AGN Rios y Acequias, Vol. 1, Exp. 9 Fa 217.
122 AGN Alcade Mayores, Vol 1, Exp. 309 Fs 439–441; AGN Tierras, Vol. 2071, Exp. 1 Fs 1–110; AGN Rios y Acequias, Vol. 1, Exp. 9 Fa 217.
123 Historia 72: Cuitzeo de la Laguna.

CHAPTER 4   RESPONDING TO CRISIS: VULNERABILITY AND ADAPTIVE CAPACITY
              IN COLONIAL MEXICO

1   Coping is usually defined as the 'manner in which people act within existing resources and range of expectations of a situation to achieve various ends' where 'resources' refers to the physical and social means of gaining a livelihood, and might include land, tools, labour, skills, crops, seed, livestock, cash and other items with value that can be sold, stored or exchanged (Blaikie et al., 1994: 62–63).
2   Adaptation can be defined as the ability of a system, individual or society to adjust to moderate potential damages, to take advantage of opportunities or to cope with the consequences.
3   Grain refers to kernels in general which can be used for consumption purposes, for animal feed or to trade. In contrast, seed is specifically for planting.
4   AHMCH Caja 2, Exp. 12.
5   AHMCH Caja 2, Exp. 12.
6   AMHCH Caja 3, Exp. 5.
7   AHMCH Notorias, Abasto de Carne, Caja 24, Exp. 10.
8   AHMCO Actas de Sesiones del Cabildo 1728–33.
9   Carga = load.
10 Juan Bautista de Fortuño, AHMCO Actas de Sesiones del Cabildo 1728–33.
11 AGN Ayuntamientos 196, Exp. 1.
12 AHMCH Guerra, Caja 2, Exp. 1.
13 AHMCH Gobierno, Caja 32, Exp. 5.
14 AHMCH Notarias, Abastos de Grano, Caja 45, Exp. 4.
15 AHMCH Hacienda, Caja 44, Exp. 1.
16 AHMCH Notarias, Abastos de Carne, Caja 45, Exp. 4.
17 AHMCH Gobierno, Caja 42, Exp. 37; AHMCH Hacienda, Caja 44, Exp. 1.
18 AGN Ayuntamientos 196 Exp. 1.
19 AHML Bandos C5 Exp. 29, 1761.
20 Léon, 1756, AGN Ayuntamientos, Vol. 196, Exp. 2.
21 AGN Caminos y Calzadas, Vol. 2, Exp. 7 Fs 131–133v.
22 AGN Ayuntamientos, Vol. 196, Exp. 4.
23 AGN Caminos y Calzadas, Vol. 2, Exp. 8 Fs 131–133v.
24 AGN Caminos y Calzadas, Vol. 2, Exp. 8 Fs 121–124.
25 AHMCH Justicia, Caja 126, Exp. 10.
26 AHMCH Demandas de Inconformidad, Caja 112, Exp. 18.

27 AHMCH Justicia, Abasto de Carne, Caja 126, Exp. 10.
28 AHMCH Gobierno, Caja 39, Exp. 24.
29 AHMCH Gobierno, Actas de Cabildo, Caja 32, Exp. 5.
30 AHMCH Gobierno, Caja 31, Exp. 9.
31 AGN Ayuntamientos 196, Exp. 4.
32 AGN Bienes de Communidad, Vol. 2, Exp. 161 Fa 225.
33 AHML Communicaciones 19.
34 AGI Patronato, Leg 226, no. 1, r. 20, Fs 3v–4v.
35 AGN Alcade Mayores, Vol. 5, Exp. 8 Fa 11.
36 Archivo de Casa Morelos, Leg 841.
37 AGN Ayuntamientos, Vol. 169, Fa 49.
38 AGN Ayuntamientos, Vol. 173, cuaderno 6.
39 AGN Ayuntamientos, Vol. 173, cuaderno 6.
40 AHMCH Justicia, Caja 126, Exp. 10.
41 AHMCH Actas de Cabildo, Caja 2, Exp. 1.
42 AMHCH Gobierno, Caja 47, Exp. 14.
43 AHMCH Actas de Cabildo, Caja 2, Exp. 1 Fs 168–169.
44 AGN Hospital de Jesus, Vol. 373, Exp. 8.
45 AHML Diezmos, Exp. 19.
46 Subject town of the cabacera of San Francisco del Rincon.
47 AGN Epidemicas, Exp. 1.
48 AGN Correspondencia Virreyes Marques de la Croix y Marques de Brancifort, Vol. 183, Fs 134–135.
49 AHMCO Actas de Sesiones del Cabildo 1728–33.
50 AHMCO Actas de Sesiones del Cabildo 1746–8, Fa 54.
51 AHMCO Actas de Sesiones del Cabildo 1746–8, Fa 78.
52 AHMCO Actas de Sesiones del Cabildo 1746–8, Fa 78.
53 AGN Correspondencia de Virreyes, Vol. 12 (Series 2), Fa 244.
54 It was not unusual for commentators to draw links between some anomalous weather events and other natural hazards. Ortlieb (1999) has highlighted how in colonial Peru contemporary populations linked earthquakes and abnormal rainfall, such that some reports on natural disasters may have led observers to erroneously attribute destruction to unusual weather rather than seismic causes. Though the case reported above deals with the implications of drought rather than heavy rains, it does illustrate the need for caution in interpreting descriptive accounts of weather and weather related events.
55 AGN Bienes Nacionales, Vol. 1182, Exp. 13.
56 The Virgin Mary of the Immaculate Conception also held appeal as a guardian against epidemic disease and other illnesses (Taylor, 1987: 19).
57 Peterson (1992) notes that rogations had already been made to several other images before Guadelupe, but that she was credited with the cessation of deaths on this occasion.
58 Porras Muñoz a la Relación de Guittierez Norriega, published in the *Boletín del AGN*, 1st Series, Vol. 30, No. 3, July-September: 370–371.
59 Archivo Arzobispado de Chihuahua, Ramo Gobierno y Administración; Cofradías, 1755, Caja 3, Series 1.3.3.

60   AGN Ayuntamientos, Vol. 173, cuaderno 3.
61   AGN Ayuntamientos, Vol. 173, cuaderno 1.
62   AHMCH Gobierno, Caja 48, Exp. 31.
63   AHMCH Gobierno, Caja 48, Exp. 31.
64   Badstue et al. (2006), however, note that there may be different points of view as to what exactly constitutes the collective and the degree to which the action reflects a common purpose. They argue that the collective can 'take different forms, ranging from formal organization to the mere observation of a set of rights and responsibilities related to the use of a common resource' (Meinzen-Dick and Di Gregorio, 2004, cited in Badstue et al., 2006).
65   AGN Tierras, Vol. 2682, Exp. 25.
66   AGN Hospital de Jesus, Leg 289, Exp. 98; AGI Audiencia de Mexico, 1684.
67   AGN Mercedes, Vol. 8, Fa 50.
68   Gay (1950), Cap. XXIX: 405.
69   AHMCH Gobierno, Caja 19, Exp. 3.
70   AHMCH Gobierno, Caja 19, Exp. 3.
71   AHMCH Hacienda, Caja 30, Exp. 10.
72   AGN Tierras, Vol. 3046, Exp. 3, Fs 74–138.
73   AGN Tierras, Vol. 3046, Exp. 3, Fs 74–138.
74   AHMCH, various.
75   AGN Tierras, Vol. 2959, Exp. 141 Fa 20v.
76   AGN Tierras, Vol. 2959, Exp. 141 Fs 10v–16v.
77   AGN Tierras, Vol. 2959, Exp. 141 Fs 4, 6v.
78   AGN Ayuntamiento, Vol. 97, Exp. 2 Fa 10.
79   Quote taken from 'Descripción de la Ciudad y Real de Minas de Guanajuato por José Hernandez Chico', 1788, Archivo de la Marina, Museo Naval de Madrid, Ms 563; in Florescano and Sanchez Gil (1975).
80   AGN Obras Publicás 17, Exp. 10 Fa 43; Descripción de la Ciudad y Real de Minas de Guanajuato por José Hernandez Chico, 1788, Archivo de la Marina, Museo Naval de Madrid, Ms 563 (Florescano and Sanchez Gil, 1975: 30).
81   AGN Obras Publicás 17, Exp. 10 Fa 43.
82   Descripción de la Ciudad y Real de Minas de Guanajuato por José Hernandez Chico, 1788, Archivo de la Marina, Museo Naval de Madrid, Ms 563 (Florescano and Sanchez Gil, 1975: 30).
83   Descripción de la Ciudad y Real de Minas de Guanajuato por José Hernandez Chico, 1788, Archivo de la Marina, Museo Naval de Madrid, Ms 563 (in Florescano and Sanchez Gil, 1975).
84   AHML Inundaciones, Exp. 3.
85   AGN Alcade Mayores, Vol. 1, Exp. 309 Fs 439–441; AGN Tierras, Vol. 2071, Exp. 1 Fs 1–110; AGN Rios y Acequias, Vol. 1, Exp. 9 Fa 217.
86   AGN Rios y Acequias, Vol. 1, Exp. 9 Fa 214.
87   AGN Rios y Acequias, Vol.1, Exp. 9 Fa 214.
88   AGN Caminos y Calzadas, Vol. 2, Exp. 8 Fs 121–124v.
89   AGN Bienes de Comunidad, Vol. 2, Exp. 295 Fs 369.
90   AGN Obras Públicas, Vol. 5, Exp. 5 Fs 249–268.
91   AGN Ayuntamientos, Vol. 180, Exp. 6 Fs 23–24.

92   AGN Ayuntamientos, Vol. 180, Exp. 6 Fs 23–24v.
93   AGN Rios y Acequias, Vol. 4, Exp. 2.
94   AGN Bienes de Communidad, Vol. 2, Exp. 295 Fa 369.
95   AGN Obras Públicas, Vol. 17, Exp. 10 Fa 43.
96   AGN Tierras, Vol. 1362, Exp. 1.
97   AGN Tierras, Vol. 1197, Exp. 2.
98   AGN Tierras, Vol. 1197, Exp 2.
99   AGN Rios y Acequias, Vol. 1, Exp. 9 Fa 214.
100 AGN Bienes de Communidad, Vol. 2, Exp. 153.
101 AGN Intendencias, Vol. 81, Exp. 1.
102 AGN Bienes de Comunidad, Vol. 2, Exp. 237 Fa 310.
103 AGN Obras Públicas, Vol. 27, Exp. 16 Fs 294–301.
104 AGN Rios y Acequias, Vol. 4, Exp. 4; AGN Tierras, Vol. 182, Exp. 1 Fa 380 (respectively).
105 AGN Tierras 1861, Exp. 7; AGN Salinas, Vol. 15, Exp. 14.
106 Relaciones Geográficas de Oaxaca in Paso y Troncoso (1939–40). Papeles de la Nueva España, 2nd Series, Vol. 4; Cuahuitlan 1580, 155–161; Chichicapa, 1580, 116–119; AGNT 450, Exp. 1; AGN Salinas 15, Exp. 14.
107 AGN Tierras, Vol. 1861, Exp. 7.
108 AGN Salinas, Vol. 15, Exp. 14; each arroba weighed 11.3 kilograms/25 pounds.
109 Conde de Revillagigedo, AGEO Alcadías Mayores, Leg 31, Exp. 14 Fa 2.
110 AGN Rios y Acequias, Vol. 4, Exp. 4.
111 AGN Tierras, Vol. 380, Exp.1 Fa 263.
112 AGN Tierras, Vol. 2979, Exp. 105 Fa 2.
113 AGN Rios y Acequias, Vol. 3.
114 AGN Tributos, 44, Exp. 15.
115 Gay (1950: ch. 19: 276).
116 AGN Tierras, Vol. 182, Exp. 1 Fa 380.
117 AGN Tierras, Vol. 182, Exp. 1 Fa 380.
118 AGN Ríos y Acequias, Vol. 1, Exp. 3.
119 AGN Tierras, Vol. 182, Exp. 1 Fa 380.
120 AGN Ríos y Acequias 1, Exp. 3.
121 AGN Inquisición, Vol. 670, Fa 201.
122 The pleito can also be traced in AGN Tierras, Vol. 675, Exp. 1; ACM, Leg 834, Fa 14.
123 AHMCH Gobierno, Caja 30, Exp. 23.
124 AHMCH Notarias, Abasto de Carnes, Caja 42, Exp. 2.
125 AHMCH Demandas de Inconformidad, Caja 60, Exp. 3.
126 AGN Alhóndigas 15, Exp. 1.
127 Chahuistle is a folk term commonly used to describe wheat rust, and less commonly used to refer to insect or worm infestations.
128 AGN Tierras, Vol. 310, Exp. 1.
129 ACM, Leg 835 (1735), Leg 838 (1700, 1706), Leg 847 (1746), Leg 860 (1711, 1718).
130 AHML Bandos, Exp. 8, 1782.
131 AHML Bandos, Exp. 8, 1782.

1 Relaciones Geográficas de Oaxaca en Paso y Troncoso (1939–40). Papeles de la Nueva España, 2nd Series, Vol. 4: Teotitlán 1579, 104–108; Nexapa 1579, 29–44.

2 Relaciones Geográficas de Oaxaca en Paso y Troncoso Franc (1939–40). Papeles de la Nueva España, 2nd Series, Vol. 4: 119–122; Amatlán; Atatlauca, 1580, 163–176.

3 Murphy (1986: 119) outlines a number of appraisals from the eighteenth century that separately value irrigated and un-irrigated land on haciendas in Celaya. Three valuations from 1739 on haciendas of comparable land are considered, one lying within the irrigation system and another two outside. The irrigated hacienda was valued at 2,400 pesos per caballería and those that were not irrigated were considered to be worth 300 pesos per caballería.

4 AGN Mercedes, Vol. 20, Fa 26; Vol. 4, Fa 303; Vol. 18, Fa 215; Vol. 18, Fa 43; Vol. 81, Fa 14; Vol. 18, Fa 8; Vol. 35, Fa 81; Vol. 18, Fa 75.

5 AGN Mercedes, Vol. 8, Fa 90.

6 For Oaxaca see, for example, AGN Indios, Vol. 5, Exp. 1017, 1591; AGN Indios, Vol. 5, Exp. 1021, 1591; AGN Indios, Vol. 11, Exp. 31, 1606; AGN Indios, Vol. 13, Exp. 146, Fs 213, 1641; AGN Indios, Vol. 31, Exp. 270, 1665; AGEO Alcadías Mayores, Leg 23, Exp. 1 Fa 9. 1755.

7 No references to this strategy were found in the examples drawn from Oaxaca, Guanajuato and Chihuahua.

8 AGN Mercedes, Vol. 8, Fa 50; Vol. 70, Fa 73; Vol. 33, Fa 32.

9 See, for example, AGN Mercedes, Vol. 73, Fa 133; Vol. 35, Fa 81; Vol. 16, Fa 287; Vol. 73, Fa 133; AGN Indios, Vol. 20, Exp. 85 Fa 1.

10 AGN Mercedes, Vol. 35, Fa 126v.

11 AGN Tierras, Vol. 674, cuad 1, Fs 100–106v.

12 AGN Tierras, Vol. 187, Exp. 2 Fs 94–108.

13 AGN Indios, Vol. 5, Exp. 154.

14 AHML Demandas, Caja 21, Exp. 10. 1590.

15 AGN Mercedes, Vol. 13, Fa 242; AGN Indios, Vol. 3, Exp. 745; AGN Indios, Vol. 5, Exp. 192; Exp. 522 all include references to the desiccation of Lake Cuitzeo at this time.

16 AGN Indios, Vol. 5, Exp. 598.

17 AGN Tierras, Vol. 187, Exp. 2 Fa 327.

18 AHMCH Justicia, Demadas de Inconformidad, Caja 21, Exp. 21.

19 Religious college/institution, often owning considerable lands and resources.

20 AGN Tierras, Vol. 149, Exp. 5.

21 AGN Tierras, Vol. 149, Exp. 5.

22 AGN Tierras, Vol. 149, Exp. 5, Fa 166.

23 AGN Tierras, Vol. 671, Exp. 3, Fa 18.

24 AGN Indois, Vol. 11, Exp. 27.

25 AGN Tierras, Vol. 41, Exp. 1 Fa 315.

26 AGEO Alcadías Mayores, Leg 20, Exp. 14 Fa 7.

27 For example, AGN Tierras, Vol. 1861, Exp. 7; Vol. 939, Exp. 1.
28 AGEO Real Intendencia Leg 54, Exp. 27 Fa 35.
29 AGN Mercedes, Vol. 70, Fa 14v.
30 AGEO Real Intendencia, Leg 1, Exp. 27.
31 AGEO Alcadías Mayores, Leg 25, Exp. 10 Fa 6.
32 AGEO Alcadías Mayores, Leg 7, Exp. 20.
33 See, for example, AGN Indios, Vol. 11, Exp. 313; AGN Mercedes, Vol. 69, Fa 50.
34 AGN Mercedes, Vol. 69, Fa 50.
35 AGN Tierras, Vol. 211, Exp. 2 Fa 48r.
36 AGN Tierras, Vol. 939, Exp. 1 Fa 48r; AGN Indios, Vol. 11, Exp. 313.
37 AGN Tierras, Vol. 353, Exp. 2 Fa 25.
38 AGN Tierras, Vol. 237, Exp. 6.
39 See, for example, AGN Tierras, Vol. 674, Exp. 1, Fa 30; follow the dispute in
    Vol. 675, Exp. 1; AGN Tierras, Vol. 192, Exp. 1; Vol. 586, Exp. 8; Vol. 1872, Exp.
    15 Fa 10; Vol. 671, Exp. 3 Fa 18; Vol. 2901, Exp. 36; Vol. 2959, Exp. 141; Vol.
    988, Exp. 1, 2 and 3 Fa 516; Vol. 1110, Exp. 18; Vol. 1166, Exp. 1 Fa 450;
    Vol. 2963, Exp. 116 Fs 246–308; Vol. 1352, Exp 1; Vol. 1368.
40 AGN Alcades Mayores, Vol. 1, Exp. 309 Fs 439–441.
41 AGN Caminos y Calzadas, Vol. 3, Exp. 1 Fs 304–312.
42 AGN Historia, Vol. 9, 1792 (Michoacán).
43 AGN Tierras, Vol. 2071, Exp. 1 Fs 1–110.
44 AGN Tierras, Vol. 1390, Exp. 3 Fa 34v, 1790.
45 AGN Tierras, Vol. 2072, Exp. 1.
46 AHMCH Gobierno, Caja 4, Exp. 8.
47 AHMCH Justicia, Demandas de Inconformidad, Caja 40, Exp. 15.
48 AHMCH Justicia, Demandas de Inconformidad, Caja 54, Exp. 1.
49 Avulsion: an abrupt change in the course of a stream or river, generally from
    one channel to a new one, often used to describe a shift in a bed of a river which
    has been used as a boundary by property owners.
50 Maps copied with kind permission from the Archivo Privado de Don Luis
    Castañeda Guzman, Oaxaca.
51 AGI Guadalajara 35 (cited in Cramaussel, 1990a: 44).
52 AGN Jesuitas, Leg 1–16, Exp. 10 Fs 10–92.
53 AHMCH Caja 2, Exp. 12.
54 AHMCH Caja 2, Exp. 12; AHMCH Caja 3, Exp. 5.
55 AHMCHI Justicia, Caja 17, Exp. 17.
56 AHMCH Porras Muñoz, Relación de Guitirrez Noriega, *AGN Boletin del AGN*
    1st Series, Vol. 30, 1959, No. 3 (July-September).
57 AHMCH Porras Muñoz, Relación de Guitirrez Noriega, *AGN Boletin del AGN*
    1st Series, Vol. 30, 1959, No. 3 (July-September).
57 AHMCH Gobierno, Caja 20, Exp. 5.
58 AGN Mercedes, Vol. 76, Fs 95–100.
59 AHMCH Guerra Caja 1, Exp. 8; Caja 1, Exp. 7.
60 AGN Mercedes, Vol. 76, Fa 137r and v.
61 AHMCH Guerra Caja 1, Exp. 7.
62 AHMCH Notarias Abastos de Carne, Caja 36, Exp. 12.
63 AHMCH Notarias Abastos de Carne, Caja 36, Exp. 12.

64 AHMCH Guerra Caja 1, Exp. 8.
65 AHMCH Guerra, Caja 1, Exp. 13.
66 AHMCH Gobierno, Caja 27, Exp. 4.
67 AHMCH Gobierno, Caja 30, Exp. 23.
68 AHMCH Guerra, Caja 2, Exp. 4.
69 AHMCH Guerra, Caja 2, Exp. 1.
70 Archivo del Ayuntamiento de Chihuahua, University of Texas at El Paso, Microfilm no. 491 (cited in Martin, 1996: 25).
71 AHMCH Notarias, Abastos de Carne, Caja 44, Exp. 20.
72 AHMCH Justicia, Demandas de Inconformidad, Caja 108, Exp. 14.
73 AGN Ayuntamientos, Vol. 173, cuaderno 3.
74 AGN Jesuitas, Leg II-9, Exp. 33.
75 AGN Jesuitas, Leg II-9, Exp. 30.
76 AGN Presidios y Carceles, Vol. 4, various fojas.
77 AHMCH Hacienda, Caja 44, Exp. 1.
78 AGN Presidios y Carceles, Vol. 4, Fs 154–160.
79 AGN Presidios y Carceles, Vol. 4, Fs 160–163.
80 Hugo O'Conor was an Irish-born officer serving the Spanish army in northern Mexico. He became Governor of the Spanish province of Texas. His role in the region was twofold. He was to bring order and discipline to the remote presidios, which had been run in a somewhat corrupt manner and to establish a new 'presidial line'. This string of strategically placed forts from the Baja to the Gulf of Mexico were intended to stamp out the Apache 'menace' once and for all.
81 AGN Presidios y Carceles, Vol. 4, Fs 205–208.
82 AGN Presidios y Carceles, Vol. 4, Fa 172.
83 AGN Presidios y Carceles, Vol. 4, Fa 172.
84 AGN Presidios y Carceles, Vol. 4, Fs 205–208.
85 AGN Presidios y Carceles, Vol. 4, Fs 340–352.
86 AHMCH Guerra, Caja 5, Exp. 5.
87 AGN Tierras, Vol. 310, Exp. 1; ACM, Leg 835 (1735), Leg 838 (1700, 1706), Leg 847 (1746), Leg 860 (1711, 1718).

CHAPTER 6   ILLUSORY PROSPERITY: ECONOMIC GROWTH AND SUBSISTENCE CRISIS
IN THE DISASTROUS EIGHTEENTH CENTURY

1 In the second half of the 1600s income from silver production from new treasuries in Pachuca and Guanajuato might have offset any revenue loss from the treasury in Mexico City (TePaske and Klein, 1981).
2 It is no coincidence that New Spain's most coveted Alcadías Mayores were those in places like Puebla, Michoacán, but particularly Oaxaca, where indigenous populations remained numerous and sizeable (Stein, 1981).
3 Alcades Mayores were unpaid and this was one of the reasons Galvez forwards to explain their acts of embezzlement and corruption. The Intendentes were to be paid 5 per cent of the annual tribute collected in their respective districts (Baskes, 2000: 45–7).

4 Galvez forwarded his proposal in 1769 when it was accepted by the Crown, but due to the opposition of the entering Mexican Viceroy, Bucareli, it would not be implemented until 1786.

5 Ponzio (2005) uses the term 'globalization' to describe this period in global history.

6 The Real Cédula was not revoked until 1809.

7 It should be remembered that many weather events may have gone unrecorded if they resulted in only minimal disruption.

8 AGN Tierras, Vol. 310, Exp. 1; ACM, Leg 835 (1735), Leg 838 (1700, 1706), Leg 847 (1746), Leg 860 (1711, 1718).

9 AGN Tierras, Vol. 149, Exp. 5, Fa 166; see chapter 5.

10 AGN Inquisición, Vol. 670, Fa 201.

11 AGN Tierras, Vol. 310, Exp. 1; ACM, Leg 835 (1735), Leg 838 (1700, 1706), Leg 847 (1746), Leg 860 (1711, 1718).

12 AGN Inquisición, Vol. 670, Fa 201.

13 Archivo Privado de Col Bustamante Vasconcelos, Papeles de San Bartolo: 94.

14 AGN Inquisición, Vol. 670, Fa 201.

15 Bazan Alarcon, A., *El Tribunal de la Acordada*, Cap VII: 51 (cited in Florescano, 1972: 91).

16 AGI Mexico 1506, 1749.

17 AHML Notarias 1750–1 Fs 144–151.

18 AHML Notarias 1750–7 Fa 24.

19 AHMCO Actas de Sessiones del Cabildo 1746–8 Fa 54.

20 AHMCO Actas de Sessiones del Cabildo 1746–8 Fa 54.

21 AGN Mercedes, Vol. 76, Fs 95–100.

22 AHMCH Notarias Abastos de Carne, Caja 36, Exp. 12.

23 AHMCH Guerra Caja 1, Exp. 7; AHMCH Notarias Abastos de Carne, Caja 36, Exp. 12.

24 AGN Mercedes, Vol. 76, Fa 137.

25 AHMCH Notarias, Abasto de Carne, Caja 36, Exp. 12.

26 AHMCH Notarias, Abastos de Carne, Caja 42, Exp. 12.

27 Archivo Arzobispado de Chihuahua, Ramo Gobierno y Administración; Cofradías, 1755, Caja 3, Serie 1.3.3.

28 AHMCH Guerra, Caja 2, Exp. 4.

29 AHMCH Gobierno, Caja 31, Exp. 14.

30 AHMCH Hacienda, Caja 49, Exp. 40.

31 AHMCH Notarias, Abastos de Carne, Caja 42, Exp. 12.

32 AGN Ayuntamientos, Vol. 173, cuaderno 3.

33 AGN Epidemias, Exp. 1.

34 AGN Jesuitas, Leg II-9, Exp. 30.

35 AGN Jesuitas, Leg II-9, Exp. 33.

36 The most severe prolonged drought is thought to have taken place between 1950 and 1965 when 14 out of 16 years had below average winter precipitation.

37 Cited in Cooper (1965: 71).

38 AGN Bandos, Vol. 13, Exp. 47 Fa 160.

39 AGN Ayuntamientos 196, Exp. 4.

40 AHML Alhóndiga Exp. 8.

41 AHML Alhóndiga Exp. 10.
42 AGN Ayuntamientos, Vol. 173, cuadernos 3.
43 *Gazeta de Mexico*, 6 December 1785, no. 52: 447–449.
44 AGN Alhóndigas, Vol. 15.
45 Pátzcuaro archives, Caja 54d. folder 1, Fs 1–150, 25 October 1785.
46 AGN Indios.
47 AHML Notorias, libro 1785, Fs 53–56.
48 AHML Notorias, libro 1786, Fs 66, 88, 98, 101, 124, 129; AHML Notorias, 1785, Fs 53–56 (the cited reference appears on Fa 55). There are a number of other references in this document to the selling off of property at 'half of the just price' (e.g. Fs 90–91, 162–3, 171–4); AHML Haciendas y Ranchos, Exp. 17; Notarias 1786.
49 AGN Ayuntamientos, Vol. 173, cuaderno 1.
50 AGN Alhóndigas, Vol. 10, Exp. 5 Fs 250–253.
51 AHML Alhóndigas, Exp. 8.
52 AHML Ayuntamientos, Vol. 169, Fa 9.
53 AGN Alhóndgas, Vol. 10, Exp. 5 Fs 250–253.
54 AHML Alhóndiga, Exp. 8.
55 AGN Historico de Hacienda, Vol. 296, Exp. 1; AHMCH Justicia, Abasto de Carne, Caja 126, Exp. 10.
56 AHMCH Guerra, Caja 5, Exp. 5.
57 AGN Historia, Vol. 72.
58 Descripción de la ciudad y Real de minas de Guanajuato por Jose Hernandez Chico, 1788. Achivo de la Marina, Museo Naval de Madrid, Ms. 563 en *Descripciones economicas regionales de la Nueva España*, 362: 19.
59 AGN Tributos, Vol. 20, Exp. 15, 2, 5.
60 AGN Correspondendia de Virreyes, Vol. 183, Fs 364–365.
61 AGN Donativos y Prestamo, Vol. 24, Exp. 29 Fs 197–204.
62 See footnote 37.
63 AGEO, Leg 4, Exp. 47 Fa 2.
64 AGN Historia, Vol. 31, no. 3.
65 AGN Obras Publicas, Vol. 41, Exp. 17.
66 AGN Obras Publicas, Vol. 41, Exp. 17.
67 A whole series of more sensational phenomena were also recorded. 'On the 22nd April of the same year (1788), in the ranch named Cachuatepeque, located on the south coast of this province, there appeared a cow-like beast with two heads, perfect in every way ... yet with two mouths ... it lived for 22 days' (AGN Historia, Vol. 31, no. 3).
68 AGN Intendencias, Vol. 61.
69 AGN Intendencias, Vol. 61.
70 AGEO, Leg 4, Exp. 47 Fa 2.
71 AGEO, Leg 4, Exp. 47 Fa 2.
72 AGEO Intendencias, Leg 1, Exp. 18.
73 AGEO Intendencias, Leg 1, Exp. 18.
74 AGN Indios, Vol. 191, Exp. 40 Fa 148.
75 AGN Epidemias 10, Exp. 1, 2, 3, 4.
76 AGN Historia 531.

77 AGN Epidemias, Vol. 4, Exp. 4.
78 Archivo Privado de Don Luis de Canstañeda.
79 AGN Rios y Acequias, Vol. 3.
80 AGN Tributos, 44, Exp. 15.
81 AGN Rios y Acequias, Vol. 1, Exp. 9 Fa 214.
82 AMHCH Gobierno, Caja 47, Exp. 14; AHMCH Notarias/Gobierno (various).
83 There had been a number of pro-insurrectionary conspiracies dating back to the 1790s and generally championed by elite criollos (Van Young, 1988).
84 It should be noted, however, that Chance and Taylor (1977) argued that growing revenue from tithes after 1784 suggests that Spaniards in Antequera were replacing Indios – who were exempt from tithe on cochineal – in the production of tithe itself.
85 AGN Civil, Vol. 302, 3 parts and 305, Fs 1763–1799.
86 AGN Industria y Comercio, Vol. 20, Exp. 6 Fa 237.
87 Rioting resumed in the 1790s, when mine operators once again sought to replace the partido with a daily wage.
88 AGN Hospital de Jesus 307, Exp. 18 (1755); AGNT 1271, Exp. 2. (1796); AGNT 1877, Exp. 2 (1797).
89 AGN Abasto y Panaderas, Vol. 2, Exp. 7 Fs 385–387.
90 AHML Caja 1809 (roll 28, INAH); Exp. 18; Caja 1809 (roll 29, INAH), Fs 4–6 (cited in Hamnett, 2002: n. 236).
91 AHMCH Hacienda Caja 55, Exp. 12.
92 AHMCH Hacienda Caja 54, Exp. 18.
93 AGN Donativos y Prestamos, Vol. 11, Exp. 26 Fs, 259–264.
94 AGN Bandos, Vol. 25, Exp. 45 Fa 62.
95 Miguel Hidalgo was an influential Creole and parish priest of Dolores (now Dolores Hidalgo in modern-day Guanajuato) and is often heralded as the father of the Mexican Independence movement.

CHAPTER 7   REGIONAL, NATIONAL AND GLOBAL DIMENSIONS OF VULNERABILITY
AND CRISIS IN COLONIAL MEXICO

1 See for example, AGEO Real Intendencia, Leg 1, Exp. 46 Fa 4; AGN Industria y Comercio, Vol. 20, Exp. 6 Fa 237; AGN Hospital de Jesus, Leg 118, Exp. 13.
2 AGN Historico de Hacienda 285 Exp. 81.
3 In Francisco del Paso y Troncoso (1939–40), Epistolario de Nueva España, 1505–1818, Vol. 4: 141–142.
4 AGN Mercedes, Vol. 81, Fa 172; AGN Mercedes, Vol. 79, Fa 17v; AGN Mercedes, Vol. 70, Fa 114; AGN Mercedes, Vol. 72, Fa 14; AGN Mercedes, Vol. 70, Fa 18vta; AGN Mercedes, Vol. 70, Fa 13; AGN Mercedes, Vol. 70, Fa 9; AGN Mercedes, Vol. 67, Fa 32; AGN Mercedes, Vol. 59, Fa 72; AGN Mercedes, Vol. 64, Fa 39; AGN Mercedes, Vol. 80, Fa 34v; AGN Mercedes, Vol. 64, Fa 109; see also Taylor (1972: 128).
5 AGEO, Leg 4, Exp. 47 Fa 2.

Bibliography

Aboites Aguilar, L. A. (1994) *Breve historia de Chihuahua*. El Colegio de Mexico, Mexico, Fondo de Cultura Economica.

Aboites Aguilar, L. A. (1995) *Norte precario. Poblamiento y colonización en Mexico (1760–1940)*. El Colegio de Mexico, Mexico, Centro de Investigaciones y Estudios Superiores en Antropología Social.

Abramovitz, J., Banuri, T., Girot, P. O., Orlando, B., Schneider, N., Spanger Siegfried, E., Switzer, J. and Hammill, A. (2001) *Adapting to climate change: natural resource management and vulnerability reduction*. Gland, Switzerland, International Union for Conservation of Nature and Natural Resources.

Acuña, R. de (ed.) (1987) *Relaciones Geográficas del siglo XVI*. Serie Antropológia, Mexico City, Universidad Nacional Autónoma de Mexico, 10 volumes.

Acuña-Soto, R., Caderon Romero, L. and Maguire, J. H. (2000) Large epidemics of haemorrhagic fevers in Mexico, *American Journal of Tropical Medicine and Hygiene* 62 (6): 733–739.

Acuña-Soto, R., Stahle, D. W., Cleaveland, M. K. and Therrell, M. D. (2002) Megadrought and megadeath in 16th century Mexico, *Emerging Infectious Diseases* 8 (4): 360–362.

Acuña-Soto, R., Stahle, D. W., Therrell, M. D., Griffin, R. D and Cleaveland, M. K. (2004) When half the population died: the epidemic of hemorrhagic fevers of 1576 in Mexico. Mini Review in *FEMS Microbiology Letters* 240: 1–5.

Acuña-Soto, R., Stahle, D. W, Cleaveland, M. K., Therrell, M. D., Chavez, S. G. and Cleaveland, M. K. (2005) Drought, epidemic disease and the fall of the classic period cultures in Mesoamerica (A 750–950). Hemorrhagic fevers as a cause of massive population loss. *Medical Hypotheses* 65 (2): 405–409.

Adger, W. N. (2003) Social capital, collective action and adaptation to climate change. *Economic Geography* 79 (4): 387–404.

Adger, W. N., Arness, N. W. and Tompkins, E. L. (2005) Successful adaptation to climate change across scales. *Global Environmental Change* 15 (2): 77–86.

Alcantara-Ayala, I. (2004) Hazard assessment of rainfall induced land sliding in Mexico. *Geomorphology* 61–2: 19–40.

Allan, R. J., Lindsay, J. and Parker, D. (1996) *El Niño/Southern Oscillation and climatic variability.* Canberra, ACT, CSIRO Publishing.

Alvarez, S. (1990) Tendencias regionales de la propiedad territorial en el norte de la Nueva España siglos XVII y XVIII. In *Actas de Segundo Congreso: Historia Regional Comparado,* Universidad Autonoma de Ciudad Juarez: 141–179.

Ambraseys, N. N. and Jackson, J. A. (1981) Earthquake hazard and vulnerability in the northeastern Mediterranean: the Corinth earthquake sequence of February–March 1981. *Disasters* 5 (4): 355–368.

Andrews, G. R. (1985) Spanish American independence: a structural analysis. *Latin American Perspectives* 12: 105–132.

Arregui, D. Lázaro de (1946) *Descripción de la Nueva Galicia,* edición y estudio de Francois Chevalier, Sevilla, Consejo Superior de Investigaciones Científicas, Escuela de Estudios Hispano Americanos.

Arrom, S. M. (1988) Popular politics in Mexico City: the Parian riot, 1828. *Hispanic American Historical Review* 68 (2): 245–268.

Bacigalupo, M. H. (1981) *A changing perspective: attitudes toward Creole society in New Spain (1521– 1610).* London, Thames Books.

Badstue L. B., Bellon, M. R., Berthaud, J., Juarez, X., Rosas, I. M, Solano, A. M, and Ramirez, A. (2006) Examining the role of collective action in an informal seed system: a case study from the central valleys of Oaxaca, Mexico. *Human Ecology* 34 (2): 249–273.

Baethgen, W. E. (1997) Vulnerability of the agricultural sector of Latin America to climate change. *Climate Research* 9: 1–7.

Ballansky, A. K. (1998) Origin and collapse of complex societies in Oaxaca (Mexico): evaluating the era from 1965 to the present. *Journal of World Prehistory* 12: 451–453.

Balkansky, A. K., Kowalewski, S. A., Perez-Rodriguez, V., Pluckahn, T. J., Smith, C. A., Stiver, L. R., Beliaev, D., Chamblee, J. F., Heredia Espinoza, V. Y. and Santos Perez, R. (2000) Archaeological survey in the Mixteca Alta of Oaxaca, Mexico. *Journal of Field Archaeology* 27 (4): 365–389.

Bancroft, H. H. (1884) History of the north Mexican states and Texas, Vol. 1 (1531–1800). In *The Works of Hubert Howe Bancroft,* Vol. 15, A.L. San Francisco, Bancroft.

Barnett, J. (2003) *The meaning of environmental security: ecological politics and policy in the new security era.* London, Zed Books.

Barrett, E. M. (1973) Encomiendas, mercedes, and haciendas in the tierra caliente of Michoacán. *Latinamerikas* 10: 71–111.

Barriendos, M. (1997) Climatic variations on the Iberian peninsula during the Late Maunder Minimum (AD 1675–1715): an analysis of data from rogation ceremonies. *The Holocene* 7: 105–111.

Barriendos, M. and Llasat, M. C. (2003) The case of the Maldá anomaly in the western Mediterranean Basin (AD 1760–1800): an example of a strong climatic variability. *Climatic Change* 61: 191–216.

Baskes, J. (1996) Coerced or voluntary? The repartimiento and market participation of peasants in late colonial Oaxaca. *Journal of Latin American Studies* 28 (1): 1–28.

Baskes, J. (2000) *Indians, merchants and markets: a reinterpretation of the repartimiento and Spanish-Indian economic relations in colonial Oaxaca, 1750–1821.* Stanford, Stanford University Press.

Baskes, J. (2005) Colonial institutions and cross-cultural trade: repartimiento credit and indigenous production of cochineal in eighteenth-century Oaxaca, Mexico. *Journal of Economic History* 65: 186–210.

Beltrán, G. A. (1946) *La población negra de Mexico, 1519–1810. Estudio etno-historico.* Mexico.

Berry, K. A. (2000) Water use and cultural conflict in 19th century North-western New Spain and Mexico. *Natural Resources Journal* 40: 759–781.

Berthe, J. P. (1970) La peste de 1643 en Michoacán. In *Historia y Sociedad en el Nuevo Mundo de habla española*, Homenaje a José Miranda, México, Mexico City, El Colegio de México: 247–261.

Berz, G., Kron, W., Loster, T., Rauch, E., Schimetschek, J., Schmieder, J., Siebert, A., Smolka, A. and Wirtz, A. (2001) World map of natural hazards – a global view of the distribution and intensity of significant exposures, *Natural Hazards* 23 (2–3): 443–465.

Beveridge, W. H. (1921) Weather and harvest cycles. *Economic Journal* 31: 429–453.

Beveridge, W. H. (1922) Wheat prices and rainfall in Western Europe. *Journal of the Royal Statistical Society*, New Series, 85: 412–478.

Blaikie, P. and Brookfield, H. (1987) *Land degradation and society.* London, Methuen.

Blaikie, P., Cannon, T., Davis, I. and Wisner, B. (1994) *At risk: natural hazards, people's vulnerability and disaster.* London, Routledge.

Blanton, R. E., Kowalewski, S. A., Feinman, G. and Appel, J. (1982) Monte Albán's hinterland, Part I: Prehispanic settlement patterns of the central and southern parts of the Valley of Oaxaca, Mexico. *Prehistory and human ecology of the Valley of Oaxaca*, Vol. 7, Memoirs of the University of Michigan Museum of Anthropology 15, Ann Arbor, Michigan.

Bogardi, J. J. (2004) Hazards, risks and vulnerabilities in a changing environment: the unexpected onslaught on human security? *Global Environmental Change* 14: 361–365.

Bohle, H. G., Downing, T. E. and Watts, M. J. (1994) Climate change and social vulnerability. *Global Environmental Change* 4 (1): 37–48.

Boissonas, B. A. (1990) *La formación de la estructura agraria en el Bajío colonial, siglos XVI y XVII.* Centro de Investigaciones y Estudios Superiores en Antropologia Social.

Borah, W. (1951) Silk raising in colonial Mexico. *Iberoamericana* 35, Berkeley, University of California Press.

Borah, W. (1955) Francisco de Urdinola's census of the Spanish settlements in Nueva Vizcaya, 1604. *Hispanic American Historical Review* 35 (3): 398–402.

Borah, W. (1992) Introduction. In Cook, N. D. and Lovell, W. G. (eds) *"Secret judgements of God:" Old World disease in colonial Spanish America.* Norman, University of Oklahoma Press: 3–19.

Borah, W. and Cook, S. F (1963) The aboriginal population of central Mexico on the eve of Spanish Conquest. *Iberoamericana* 45. Berkeley, University of California Press.

Boyer, R. (1977) Mexico in the seventeenth century: transition of a colonial society. *Hispanic American Historical Review* 57 (3): 455–478.

Bradbury, J. P. (1989) Late Quaternary lacustrine palaeoenvironments in the Cuenca de Mexico. *Quaternary Science Reviews* 8: 75–100.

Brading, D. A. (1971) *Miners and merchants in Bourbon Mexico, 1730–1810.* Cambridge, Cambridge University Press.

Brading, D. A. (1973) Government and elite in late colonial Mexico. *Hispanic American Historical Review* 53 (3): 389–414.

Brading, D. (1978) *Haciendas and ranches in the Mexican Bajio, Leon, 1700–1860.* Cambridge, Cambridge University Press.

Bradley, R. S. (1999) *Paleoclimatology: reconstructing climates of the Quaternary.* San Diego, Harcourt Academic Press.

Brázdil, R., Valášek, H., and Macková, J. (2003) Climate in the Czech Lands during the 1780s in light of the daily weather records of parson Karel Bernard Hein of Hodonice (southwestern Moravia): comparison of documentary and instrumental data. *Climatic Change* 60: 297–327.

Brazdil, R., Pfister, C., Wanner, H., von Storch, H. and Luterbacher, J. (2005) Historical climatology in Europe – the state of the art. *Climatic Change* 70 (3): 363–430.

Brown, N. and Isaar, A. (eds) (1999) *Water, environment and society in times of climatic change.* Dordrecht, Kluwer Academic Press.

Burkholder, M. A. and Johnson, L. L. (1994) Colonial Latin America. Oxford, Oxford University Press.

Burningham, K. and Cooper, G. (1999) Being constructive: social constructionism and the environment. *Sociology* 33 (2): 297–316.

Burroughs, W. J. (2001) *Climate change: a multidisciplinary approach.* Cambridge, Cambridge University Press.

Burton, I., Kates, R. W. and White, G. F. (1993) *The environment as hazard.* New York, Guildford Press.

Bustamante, J. I. (1992) *Temas del pasado Oaxaqueño.* Fundacion Cultural Bustamante Vascincellos.

Butzer, E. K. (2003) *Floods, epidemic and human response in Northeastern Mexico, 1802.* Paper presented at the Annual Conference of the Association of American Geographers, New Orleans, March.

Butzer, K. W. (1990) Ethno-agriculture and cultural ecology in Mexico: historical vistas and modern implications. *Benchmark 1990, Conference of Latin Americanist Geographers* 17/18: 139–153.

Butzer, K. W. (1991) Spanish colonization of the New World: cultural continuity and change in Mexico. *Erdkunde* 45: 205–219.

Butzer, K. W. (2005) Environmental history in the Mediterranean world: cross-disciplinary investigation of cause-and-effect for degradation and soil erosion. *Journal of Archaeological Science* 32: 1773–1800.

Butzer, K. W. and Butzer, E. K. (1993) The sixteenth century environment of the central Mexican Bajío: archival reconstruction from colonial land grants and the questions of Spanish ecological impact. In Matthewson, K. (ed.) *Culture, form and place: essays in cultural and historical geography.* Geosceince and Man, Vol. 32: 89–124.

Butzer, K. W. and Butzer, E. K. (1995) Transfer of the Mediterranean livestock economy to New Spain: adaptation and ecological consequences. In Turner, B. L. and Sal, G. A. (eds) *Global land use change.* Mexico, Consejo Superior de Investigaciones Cientificas: 151–193.

Butzer, K. W. and Butzer, E. K. (1997) The natural vegetation of the Mexican Bajío: archival documentation of a sixteenth century savannah environment. *Quaternary International* 43/44: 161–172.

Camuffo, D. and Enzi, S. (1992) Reconstructing the climate of northern Italy from archive sources. In Bradley, R. S. and Joes, P. D. (eds) *Climate since AD 1500*. London, Routledge: 143–154.

Carter, H. R. (1931) *Yellow fever: an epidemiological and historical study of its place of origin*. Baltimore.

Castiglia, P. J. and Fawcett, P. J. (2006) Large Holocene lakes and climate change in the Chihuahuan Desert. *Geology* 34 (2): 113–116.

Cavazos, T. and Hastenrath, S. (1990) Convection and rainfall over Mexico and their modulation by the Southern Oscillation. *International Journal of Climatology* 10: 377–386.

Cavo, A. P. (1949) *Historia de México*, anotada por Ernesto Burrus, prologo Mariano Cuevas, México, Editorial Patria.

Chance, J. K. (1976) The urban Indian in colonial Oaxaca. *American Ethnologist* 3 (4): 603–632.

Chance, J. K. and Taylor, W. B. (1977) Estate and class in a colonial city: Oaxaca in 1792. *Comparative Studies in Society and History* 19 (4): 454–487.

Changnon, S. A., Pielke, R. A., Jr., Changnon, D., Sylves, R. T. and Pulwarty, R. (2000) Human factors explain the increased losses from weather and climate extremes. *Bulletin of the American Meteorological Society* 81 (3): 437–442.

Chaunu, H. and Chaunu, P. (1955) *Seville et l'atlantique, 1504–1650*. Paris, Centre de Recherches Historiques.

Chen, R. S. and Kates, R. W. (1994) World food security: prospects and trends. *Food Policy* 19 (2): 192–208.

Chenoweth, M. (1998) The early 19th century climate of the Bahamas and a comparison with 20th century averages. *Climatic Change* 40 (3–4): 577– 603.

Chenoweth, M. (2005) *The eighteenth century climate of Jamaica derived from the journals of Thomas Thistlewood, 1750–1786*. Transactions of the American Philosophical Society Series. American Philosophical Society.

Chevalier, F. (1952) *La formación des grandes domaines au Mexique: terre et societé aux XVI–XVII siécles*. Paris, Institut d'Ethnologie.

Christensen, A. F. (1998) The 1737 Matlazahuatl epidemic in Mixquiahuala and Tecpatepec Mexico. *AAPA Abstracts* 76–77.

Ciudad Real (1976) Descripción de la Ciudad y Real de Minas de Guanajuato por José Hernandez Chico, 1788. Archivo de la Marina, Museo Naval de Madrid, Ms 563. In Florescano, E. and Gil, I. (eds) CPC2 *Descripciones económicas regionales de la Nueva España 1766–1827*. Fuentes para la historia economica III. Mexico, SE-INAH.

Clark, W. C. et al. (2000) Assessing vulnerability to global environmental risks. Report of the Workshop on Vulnerability to Global Environmental Change: Challenges for Research, Assessment and Decision Making. Research and Asssessment Systems for Sustainability Porgram Discussion Paper 2000–12. Warrenton, VA: Environment and Natural Resources Program, Belfer Centre for Science and International affairs (BCSIA), Kennedy School of Government.

Claxton, R. H and Hecht, A. D. (1978) Climatic and human history in Europe and Latin America. *Climatic Change* 1: 195–203.

Cleaveland, M. K., Stahle, D. W., Therrell, M. D., Villanueva-Diaz, J. and Burns, B. T. (2003) Tree-ring reconstructed winter precipitation and tropical teleconnections in Durango, Mexico. *Climatic Change* 59: 369–388.

Cobb, R. (1970) *The police and the people: French popular protest, 1789–1820.* Oxford, Oxford University Press.

Coello, J. (1989) The persistence of indian slavery and encomienda in the northeast of colonial Mexico, 1577–1723. *Journal of Social History* 21: 683–700.

Conde, C., Liverman, D., Flores, M., Ferrer, R., Araújo, R., Betancourt, E., Villarreall, G. and Gay, C. (1997) Vulnerability of rainfed maize crops in Mexico to climate change. *Climate Research* 9: 17–23.

Conteras, S. C. (1999) *El clima de la República Mexicana en el siglo XIX, Mexico.* Doctoral thesis. Mexico, Facultad de Filosofia y Letras, UNAM.

Cook, S. F. (1949) *Soil erosion and population in central Mexico.* Berkeley, University of California Press.

Cook, S. F. and Borah, W. (1957) The rate of population change in central Mexico. *Hispanic American Historical Review* 37: 463–470.

Cook, S. F. and Borah, W. (1960) The Indian population of central Mexico, 1531–1610. *Iberoamericana* 44. Berkeley, University of California Press.

Cook, S. F. and Borah, W. (1968) The population of the Mixteca Alta. *Iberoamericana* 50. Berkeley, University of California Press.

Cook, S. F. and Borah, W. (1971, 1974, 1979) *Essays in population history,* 3 vols. Berkeley, University of California Press.

Cook, S. F. and Simpson, C. B. (1948) The population of central Mexico in the sixteenth century. *Iberoamericana* 31. Berkeley, University of California Press.

Coombes, P. and Barber, K. (2005) Environmental determinism in Holocene research: casualty or coincidence? *Area* 37 (3): 303–311.

Cooper, D. (1965) *Epidemic disease in Mexico City, 1761–1813.* Institute of Latin American Studies, Austin, University of Texas Press.

Cope, R. D. (1994) *Limits of racial domination: plebeian society in colonial Mexico City, 1660–1720.* Madison, University of Wisconsin Press.

de Cossio, F. G. (1856) *Legislación indigenista de Mexico.* Interamericano, Ediciones Speciales, no. 38, Mexico: 39–44.

Cramaussel, C. (1990a) *La provincia de Santa Barbara en Nueva Vizcaya 1563–1631.* Chihuahua, Universidad Autonoma de Cuidad Juarez.

Cramaussel, C. (1990b) Evolucion de las formas de dominio del espacio colonial las haciendas de la region de Parral. In *Actas del segundo congreso: historia regional comparado.* Chihuahua, Universidad Autonoma de Ciudad Juarez: 115–140.

Crosby, A. (1979) Virgin soil epidemics as a factor in the aboriginal depopulation in America. *William and Mary Quarterly* 33: 289–299.

Crosby, A. (1986) *Ecological imperialism: the biological expansion of Europe 900–1900.* New York, Canto.

Curtin C. G., Sayre, N. F. and Lane, B. D. (2002) Transformations of the Chihuahuan borderlands: grazing, fragmentation and biodiversity conservation in desert grasslands. *Environmental Science and Policy* 5: 55–68.

Curtis, J. H., Hodell, D. A. and Brenner, M. (1996) Climate variability on the Yucatan peninsula (Mexico) during the past 3500 years and implications for Maya cultural evolution. *Quaternary Research* 56 (1): 37–47.

Curtis, J. H., Brenner, M., Hodell, D. A., Balser, R. A., Islebe, G. A. and Hooghiemstra, H. (1998) A multiproxy study of Holocene environmental change in the Maya lowlands of Peten, Guatemala. *Journal of Palaeolimnology* 19: 139–159.

Cutter, S. L. (1996) Vulnerability to environmental hazards. *Progress in Human Geography* 20 (4): 529–539.

Cutter, S. L. (ed.) (2006) *Hazards, vulnerability and environmental justice*. London, Earthscan.

Daanish, M. (2002) Linking access and vulnerability: perceptions of irrigation and flood management in Pakistan. *Professional Geographer* 54 (1): 94.

Davidson, N. (2004) The Scottish path to capitalist agriculture: from the crisis of feudalism to the origin of agrarian transformation (1688–1746). *Journal of Agrarian Change* 4 (3): 227–268.

Deeds, S. M. (1989) Rural work in Nueva Vizcaya: forms of labor coercion on the periphery. *Hispanic American Historical Review* 69 (3): 425–449.

Deeds, S. M. (2003) *Defiance and deference in Mexico's colonial north*. Austin, University of Texas Press.

Demarée, G. R., Ogilvie, A. E. J. and Zhang, D. (1996) Further documentary evidence of northern hemispheric coverage of the great dry fog of 1783. Comment on Stothers, R. B. "The great dry fog of 1783" (published in *Climatic Change* 32). *Climatic Change* 39: 727–730.

Demeritt, D. (2001) Being constructive about nature. In Castree, N. and Braun, B. (eds) *Social nature, theory, practice and politics*. Oxford, Blackwell: 22–40.

Dettinger, M. D., Cayan, D. R., Diaz, H. F. and Meko, D. (1998) North-South precipitation patterns in Western North America on interannual to decadal timescales. *Journal of Climate* 11: 3095–3111.

De Vries, J. (1980) Measuring the impact of climate on history: the search for appropriate methodologies. *Journal of Interdisciplinary History* 10: 599–630.

Dias, S. C., Therrell, M. D., Stahle, D. W. and Cleaveland, M. K. (2002) Chihuahua (Mexico) winter-spring precipitation reconstructed from tree-rings, 1647–1992. *Climate Research* 22: 237–244.

Dilley, M. (1997) Climatic factors affecting annual maize yields in the Valley of Oaxaca, Mexico. *International Journal of Climatology* 17: 1549–1557.

Donkin, R. A. (1977) Spanish red: an ethnographical study of cochineal and the opuntia cactus. *Transactions of the American Philosophical Society* 67 (5): 3–84.

Doolittle, W. E. (1984) Settlements and the development of "statelets" in Sonora, Mexico. *Journal of Field Archaeology* 11 (1): 13–24.

Doolittle, W. E. (1989) Pocitos and registros: comments on water-control features at Hierve el Agua, Oaxaca. *American Antiquity* 54 (4): 841–847.

Douglas, M. (1985) *Risk acceptability according to the social sciences*. New York, Russell Sage Foundation.

Douglas, M. and Wildavsky, A. (1982) How can we know the risks we face? Why risk selection is a social process. *Risk Analysis* 2: 49–51.

Dunn, M. M. (1988) *Political essay on the Kingdom of New Spain by Alexander Von Humboldt*. The John Black Riley Translation, New York (abridged). Ed. M. Maples Dunn. Norman, University of Oklahoma Press.

Eakin, H. (2005) Institutional change, climate risk and rural vulnerability: cases from central Mexico. *World Development* 33 (11): 1923–1938.

Easterling, D. R., Evans, J. L., Groisman, P. Y., Karl, T. R., Kunkel, K. E. and Ambenje, P. (2001) Observed variability and trends in extreme climate events: a brief review. *Bulletin of the American Meteorological Society* 81 (3): 417–425.

Endfield, G. H. (1997) *Social and environmental change in colonial Michoacán, west central Mexico.* Unpublished Ph.D. Thesis, University of Sheffield.

Endfield, G. H. (1998) Lands, livestock and the law: territorial conflicts in colonial wets central Mexico. *Colonial Latin American Review* 7 (2): 205–224.

Endfield, G. H. and Fernández-Tejedo, I. (2006) Decades of drought, years of hunger: archival investigations of multiple year droughts in late colonial Chihuahua. *Climatic Change* 75 (4): 391–419.

Endfield, G. H. and O'Hara, S. L. (1997) Conflicts over water in the "Little Drought Age" in central Mexico. *Environment and History* 3: 255–272.

Endfield, G. H and O'Hara, S. L (1999) Degradation, drought and dissent: an environmental history of colonial Michoacán, west central Mexico. *Annals of the Association of American Geographers* 89 (3): 402–419.

Endfield, G. H., Fernández-Tejedo, I. and O'Hara, S. L (2004a) Conflict and cooperation: water, floods and social response in colonial Guanajuato, Mexico. *Environmental History* 9 (2): 221–247.

Endfield G. H., Fernández-Tejedo, I. and O'Hara, S. L. (2004b) Drought and disputes, deluge and dearth: climatic variability and human response in colonial Oaxaca, Mexico. *Journal of Historical Geography* 30: 249–276.

Epstein, P. R. (1999) Climate and health. *Science* 285: 347–348.

Epstein, P. R. (2001) Climate change and emerging infectious diseases. *Microbes and Infection* 3: 747–754.

Epstein, P. R. (2002) Climate change and infectious disease: stormy weather ahead? *Epidemiology* 13 (4): 272–375.

Epstein, P. R., Dobson, A. and Vendemeer, J. (1997) Biodiversity and emerging infectious diseases: integrating health and ecosystem monitoring. In Griffo, F. and Rosenthal, J. (eds) *Biodoversity and health.* Washington, DC, Island Press.

Fabila, M. (1941) *Cinco siglos de legislación agraria (1493–1940).* Mexico.

Feinman, G. M., Kowalewski, S. A., Finsten, L., Blanton, R. E. and Nicholas, L. (1985) Long-term demographic change: a perspective from the Valley of Oaxaca. *Journal of Field Archaeology* 12 (3): 333–362.

Fernández-Tejedo, I., Endfield, G. H. and O'Hara, S. L. (2004) Estratagias para el control del agua en Oaxaca colonial. *Estudios de Historia Novohispana* 31: 137–198.

Fisher, J. R. (1997) *The economic aspects of Spanish imperialism in America, 1492–1820.* Liverpool, Liverpool University Press.

Flannery, K. V. (ed.) (1969) *Preliminary archaeological investigations in the Valley of Oaxaca, Mexico, 1966–69. A Report to the National Science Foundation and the Instituto Nacional de Antropologia e Historia.* Mimeograph. Ann Arbor, University of Michigan Museum of Anthropology.

Flannery, K. V. and Spores, R. (1983) Excavated sites of the Oaxaca pre-ceramic. In Flannery, K. V. and Marcus, J. (eds) *The Cloud People: divergent evolution of the Zapotec and Mixtec civilizations.* New York, Academic Press: 20–26.

Flannery, K. V., Marcus, J. and Kowalewski, S. A. (1981) The Preceramic and Formative of the Valley of Oaxaca. In Sabloff, J. A. (ed.) *Supplement of the Handbook of Middle American Indians, Volume 1: Archaeology*: 48–93.

Florescano, E. (ed.) (1972) *Analysis historico de las sequías en Mexico, 1708–1810.* Mexico City, Colegio de Mexico.

Florescano, E. (1976) *Origen y desarollo de los problemas agrarios de Mexico (1500–1821).* Colección Problemas de Mexico. Mexico, Ediciones Era.

Florescano, E. (1980) Una historia olvidada: la sequia en Mexico. *Nexos* 32: 9–13.

Forescano, E. (1981) *Fuentes para la historia de la crisis agrícola de 1785–1786.* Mexico City, Archivo General de la Nación.

Florescano, E. (1986) *Precios del maiz y crisis agricolas en Mexico: 1708–1810.* Mexico, Ediciones Era.

Florescano, E. and Malvido, E. (1982) *Ensayos sobre la historia de las epidemias en Mexico.* Mexico, Colección Salud y Seguridad Social, Serie Historia. Tomo 1.

Florescano, E. and Sanchez Gil, I. (eds) (1975) *Descripciones economicas regionales de la Nueva España 1766–1827.* Mexico, SE-INAH. Fuentes para la historia economica.

Florescano, E. and Sanchez Gil, I. (2000) La epoca de las reformas borbonicas y el crecimiento economico, 1750–1808. In *Historia General de Mexico,* Mexico, El Colegio de Mexico: 471–589.

Florescano, E., Swan, S., Menegus, M. and Galindo, I. (1995) *Breve historia de la sequia en Mexico.* Veracruz, Universidad Veracruzana.

Foster, H. (1980) *Disaster planning: the preservation of life and property.* New York, Springer.

Fraser, E. D. G., Mabee, W. and Slaymaker, O. (2003) Mutual vulnerability, mutual dependence: the reflexive relation between human society and the environment. *Global Environmental Change* 13: 137–144.

Frazier, K. (1979) *The violent face of nature.* New York, William Morrow.

Frizzi, R. M. de los Angeles (1991) Introducción. Oaxaca: 1786–1876. In Lecturas históricas del estado de Oaxaca, v.iii, siglo XIX, Mexico, Instituto Nacional de Antropología e Historia: 15–50.

Gabor, T. and Griffith, T. K. (1980) The assessment of community vulnerability to acute hazardous materials incidents. *Journal of Hazardous Materials* 8: 323–333.

García, E. (1974) Situaciones climaticos durante el auge y la caida de la cultura Teotihuacana. *Boletín del Instituto de Geografia,* 5. Mexico City, Universidad Nacional Autónoma de Mexico: 35–69.

García-Herrera, R., Garcia, R. R., Prieto, R. M del, Hernandez, E., Gimeo, L. and Diaz, H. F. (2003) The use of Spanish historical archives to reconstruct climate variability. *Bulletin of the American Meteorological Society* 4 (8): 1025–1035.

Garcia, S. J. (2000) Analysis de vulnerabilidad agropecuaria por sequia en el estado de Chihuahua. Instituto de Ecologia, Jalapa, Veracruz; available at www.sequia.edu.mx/proyectos/vulnera.html.

García-Acosta, V. (1993) Las sequias historicas de Mexico. *La Red* 1: 2–18.

García-Acosta, V. (ed.) (1997) *Historia y desastres in America Latina. Red de Estudios Sociales en prevención de desastres en América Latina.* Mexico, CIESAS.

Garcia, R., Díaz, H. F., García Herrera, R., Eischeid, J., del R. Prieto, M., Hernández, E., Gimeno, L., Durán, F. R. and Bascary, A. M. (2001) Atmospheric circulation changes in the tropical Pacific inferred from the voyages of the Manila galleons in the sixteenth to eighteenth centuries. *Bulletin of the American Meteorological Society* 82 (11): 2435–2455.

Garcia-Herrera, R., Gemeno, L., Ribera, P. and Hernández, E. (2005) New records of Atlantic hurricanes from Spanish documentary sources. *Journal of Geophysical Research* 110.

Gay, J. A. (1950) *Historia de Oaxaca*, 4 vols, 3rd edn. México, Editorial Porrúa.

Gerhard, P. (1982). *The north frontier of New Spain*. Princeton, Princeton University Press.

Gibson, C. (1964) *The Aztecs under Spanish rule*. Stanford, Stanford University Press.

Gibson, C. (1982) The Aztecs under Spanish rule: a history of the Indians of the Valley of Mexico. Stanford: Stanford, University Press.

Gioda, A. and L'Hôte, Y. (2002) Archives, histoire du clima et pluviométrie: un exemple Sud-Américain. *La Houille Blanche* 4 (5): 44–50.

Gomez, A. and Armando, A. (1995) *El comercio interno de la Nueva España, el basto de la ciudad de Guanajuato, 1770–1810*. Mexico, INAH.

Gonzalez, P. (1904) *Geografía local del estado de Guanajuato*, Tip. De La Escuale Ind Militar.

Gradie, C. M. (2000) *The Tepehuan revolt of 1616: militarism, evangelism and colonialism in seventeenth century Nueva Vizcaya*. Salt Lake City, University of Utah Press.

Grattan, J. and Brayshay, M. (1995) "An amazing and portentious summer": environmental and social responses in Britain to the 1783 eruption of an Iceland volcano. *Geographical Journal* 161 (2): 125–134.

Griffen, W. B. (1979) *Indian assimilation in the Franciscan area of Nueva Vizcaya*. Tuscon, University of Arizona Press.

Grove, J. M. (1983) Tax records from west Norway as an index of the Little Ice Age. *Climatic Change* 5: 265–282.

Grove, R. (1998) Global impact of the 1789–1793 El Niño. *Nature* 393: 318–319.

Gunder Frank, A. (1969) The development of under development. In Gunder Frank, A. (ed.) *Latin America: under development or revolution*. New York, Monthly Review Press: 3–17.

Hammond, G. P. and Rey, A. (eds and trans.) (1927) *Don Juan de Oñate: colonizer of New Mexico, 1595–1628*, 2 vols. Albuquerque, University of New Mexico Press.

Hamnett, B. R. (1969) The appropriation of Mexican church wealth by the Spanish Bourbon government – the "Consolidation" de Vales Reales, 1805–1809. *Journal of Latin American Studies* 1 (2): 85–113.

Hamnett, B. R. (1971). *Politics and trade in southern Mexico 1750–1821*. Cambridge, Cambridge University Press.

Hamnett, B. R. (1985) *La politica española en una epoca revolucionaria, 1790–1820*. Mexico City.

Hamnett, B. R. (1997) Process and pattern: a re-examination of the Ibero-American independence movements, 1808–1826. *Journal of Latin American Studies* 29 (2): 279–328.

Hamnett, B. R. (2002) *Roots of insurgency: Mexican regions, 1750–1824*. Cambridge Latin American Studies. Cambridge, Cambridge University Press.

Hassan, F. (2000) Environmental perception and human responses in history and prehistory. In McIntosh, R. J., Tainter, J. A. and McIntosh, S. K. (eds) *The way the wind blows: climate, history and human action*. New York, Columbia University Press: 121–140.

Hassig, R. (1981) The famine of one rabbit: ecological causes and social consequences of a pre-Columbian calamity. *Journal of Anthropological Research* 37: 172–182.

Hastenrath, S. (1988) *Climate and circulation of the Tropics*. Boston, D. Reidel.

Heine, K. (1988) Late Quaternary glacial chronology of the Mexican volcanoes. *Die Geowissenschaften* 6: 197–205.

Herr, R. (1958) *The eighteenth century revolution in Spain*. Princeton, Princeton University Press.

Hewitt, K. and Burton, I. (1971) *The hazardousness of a place: a regional ecology of damaging events*. Toronto, University of Toronto Press.

Hewitt, W. P., Winter, M. C. and Peterson, D. A. (1987) Salt production at Hierve el Agua, Oaxaca. *American Antiquity* 52: 799–816.

Higgins, R. W., Chen, Y. and Douglas, A. V. (1999) Interannual variability of the North American warm season precipitation regime. *Journal of Climate* 12: 653–680.

Hobsbawm, E. and Rudé, G. (1968) *Captain swing: a social history of the great English agricultural uprising of 1830*. New York, Pantheon Press.

Hocquenghem, A. and Ortlieb, L. (1992) Historical records of El Niño events in Peru (XVI-XVIIIth centuries: the Quinn *et al* (1987) chronology revisited. In Ortlieb, L. and Macharé, J. (eds) *Paleo ENSO Records International Symposium Extended Abstracts*. Lima, Peru, OSTROM/CONCYTEC: 143–149.

Hodell, D. A., Brenner, M. and Curtis, J. H. (2005a) Terminal Classic drought in the northern Maya lowlands inferred from multiple sediment cores in Lake Chichancanab. *Quaternary Science Reviews* 24: 1413–1427.

Hodell, D. A., Brenner, M., Curtis, J. H., Medina-González, R., Idelfonso Chan Can, E., Abornaz-Pat, A. and Guilderson, T. P. (2005b) Climate change on the Yucatan Peninsula during the Little Ice Age. *Quaternary Research* 63: 109–121.

Hodell, D. A., Curtis, J. H. and Brenner, M. (1995) Possible role of climate in the collapse of the Classic Maya Civilization. *Nature* 375: 391–394.

Homer-Dixon, T. (1991) On the threshold: environmental changes as causes of acute conflict. *International Security* 16 (2): 76–116.

Homer-Dixon, T. (1995) Environmental scarcities and violent conflict: evidence from cases. *International Security* 19 (1): 5–40.

Houghton, J. T., Ding, Y., Griggs, D. J., Noguer, M., van der Linden, P. J., Dai, X., Maskell, K. and Johnson, C. A. (2001) *Climatic change 2001: the scientific basis*. Geneva, Switzerland, Intergovernmental Panel on Climate Change.

Humboldt, A. (1811) (1973 reprint) *Ensayo politico sobre el reino de la Nueva España*, estudio preliminar y notas de Juan A. Ortega y Medina. Mexico, Editorial Porrua.

Ingram, M. J., Farmer, G. and Wigley, T. M. L. (1981) Past climates and their impact on man: a review. In Wigley, T. M. L. Ingram, M. J and Farmer, G. (eds) *Climate and history*. Cambridge, Cambridge University Press.

Instituto Nacional de Estadística, Geografica, e Informatica (INEGI) (accessed 2003) http://www.inegi.gob.mx/geo/default.asp?e=08&c=124.

Intergovernmental Panel on Climate Change (2001) *Climate change 2001, third assessment. Working Group II report: impacts, adaptation and vulnerability*. IPCC. GRID Arendal.

Intergovernmental Panel on Climate Change (2007) *Climate change 2007, fourth assessment. Working Group II report: impacts, adaptation and vulnerability*. Geneva, Switzerland, IPCC.

International Strategy for Disaster Reduction (2002) *Drought. Living with risk: an integrated approach to reducing societal vulnerability to drought.* ISDR Ad Hoc Discussion Group on Drought, Report.

Israel, J. I. (1974) Mexico and the "general crisis" of the seventeenth century. *Past and Present* 63: 33–57.

Israel, J. I. (1975) *Race, class and politics in colonial Mexico, 1610–1670.* Oxford, Oxford University Press.

Jáuregui, E. (1979) Algunos aspectos de las fluctuaciones pluviométricas en Mexico en los últimos cien años. *Boletín del Instituto de Geografía* 9: 39–64.

Jáuregui, E. (1997) Climatic changes in Mexico during the historical and instrumented periods. *Quaternary International* 43/44: 7–17.

Jáuregui, E. and Kraus, D. (1976) Some aspects of climate fluctuations in Mexico in relation to drought. *Geofísica Internacional* 16: 45–61.

Jones, P. D. (1988) It was the best of times, it was the worst of times. *Science* 280 (5363): 544–545.

Jones, P. D., Jonsson, T. and Wheeler, D. (1997) Extension to the North Atlantic Oscillation using early instrumental pressure observations from Gibraltar and southwest Iceland. *International Journal of Climatology* 17: 1433–1450.

Jones, S. (2002) Social constructionism and the environment: through the quagmire. *Global Environmental Change* 12 (4): 247–251.

Joralemon, D. (1982) New World depopulation and the case of disease. *Journal of Anthropological Research* 38 (1): 108–127.

Joyce, A. A. and Mueller, R. G. (1997) Prehispanic human ecology of the Rio Verde drainage basin, Mexico. *World Archaeology* 29 (1): 75–94.

Kamen, H. (1991) *Spain, 1469–1714: a society of conflict.* London, Longman.

Kasperson, J. X., Kasperson, R. E. and Turner III, B. L. (1995) *Regions at risk: comparisons of threatened environments.* Tokyo, United Nations University Press.

Kasperson, R. E. and Kasperson, J. X. (1996) The social amplification and attenuation of risk. *Annals of the Academy of Social and Political Science* 545: 95–105.

Kasperson, R. E. and Kasperson, J. X. (2001) *Global environmental risk.* London, Earthscan.

Kasperson, R. E., Renn, O., Solvic, P., Brown, H. S., Emel, J., Goble, R., Kasperson, J. X. and Ratick, S. (1988) The social amplification of risk: a conceptual framework. *Risk Analysis* 8 (2): 177–187.

Katz, F. (1988) Rural uprisings in pre-Conquest and colonial Mexico. In *Riot, rebellion and revolution: rural social conflict in Mexico.* Princeton, Princeton University Press: 65–94.

Kelley, J. C. (1951) A Bravo Valley aspect component of the lower Rio Conchos Valley, Chihuahua, Mexico. *American Antiquity* 17 (2): 114–119.

Kelley, J. C. (1992) La Cuenca del Río Conchos, historia, arquaeología y significado. In Marquez-Alameda, A. (ed.) *Historia General de Chihuahua I: Geología, Geografía y Arqueología.* Universidad Autónoma de Cuidad Juarez. Gobierno del Estado de Chihuahua.

Kim, T.-W., Valdez, J. B. and Aparicio, J. (2002) Frequency and spatial characteristics of droughts in the Conchos River Basin, Mexico. *Water International* 27 (3): 420–430.

Kington, J. (1980) Daily weather mapping from 1781. A detailed synoptic examination of weather and climate change during the decade leading up to the French revolution. *Climatic Change* 3: 7–36.

Kington, J. (1988) *The weather of the 1780s over Europe*. Cambridge, Cambridge University Press.

Kirkby, A. V. T. (1973) The use of land and water resources in the past and present Valley of Oaxaca, Mexico. *Memoirs of the Museum of Anthropology* 5. Ann Arbor, University of Michigan.

Kirkby, A. V. T (1974) Individual and community responses to rainfall variability in Oaxaca, Mexico. In G. F. White, *Natural hazards: local, national, global*. Oxford, Oxford University Press: 119–128.

Knight, A. (2002) *Mexico: the colonial era*. Cambridge, Cambridge University Press.

Kohle, U. S. de M. and Dandekar, M. M. (2004) Natural hazards associated with meteorological extreme events. *Natural Hazards* 31: 487–497.

Koselleck, R. (1985) Futures past: on the semantics of historical time. Trans. K. Tribe. Cambridge, MA, MIT Press.

Kovats, R. S., Bouma, M. J., Hajat, S., Worrall, E. and Haines, A. (2003) El Niño and health. *The Lancet* 362 (9394): 1481–1489.

Kowalewski, S. A., Feinman, G. M., Finsten, L., Blanton, R. E. and Nichoals, L. M. (1989) Monte Alban's hinterland, Part II. Prehispanic settlement patterns in Tlacolula, Etla and Ocotlán, the Valley of Oaxaca, Mexico. *Memoirs of the Museum of Anthropology* 23. Ann Arbor, University of Michigan.

Kundzewicz, Z. W. and Kaczmarek, Z. (2000) Coping with hydrological extremes. *Water International* 25: 66–75.

Kunkel, K. E., Pielke, R. A., Jr. and Cahngnon, S. A. (1999) Temporal fluctuations in weather and climate extremes that cause economic and human health impacts. A review. *Bulletin of the American Meteorological Society* 80 (6): 1077–1098.

Kutzbach, J. E. and Street-Perrott, E. (1985) Milankovitch forcing of fluctuations in the level of Tropical lakes from 18 to 0 kyr BP. *Nature* 317: 130–134.

Laferrier, J. E. (1992) Cultural and environmental response to drought among the mountain Pima. *Ecology of Food and Nutrition* 28 (1–2): 1–9.

Lafora, N. (1939) *Relación de Viaje que hizo a los presidios internos en la frontera de la America septentrional*. Mexico.

Lafora, N. (1958) Viaje a los presidios internos de la America Septrional. Estudio preliminar y notas por Mario Hernández y Sanchez-Barba. In *Viajes y Viajeros, Viajes por Norteamerica*. Madrid, Aguilar: 258–327.

Lamb, H. H. (1965) The early medieval warm epoch and its sequel. *Palaeogeography, Palaeoclimatology and Palaeoecology* 1: 13–37.

Landsberg, H. E. (1980) Past climates from unexploited written sources. *Journal of Interdisciplinary History* 10 (4): 631–642.

Le Roy Ladurie, E. (1983) *Historie du climat depuis l'an mil*. Paris, Flammarion.

Licate, J. A. (1981) *Creation of a Mexican landscape: territorial organization and settlement in the Eastern Puebla Basin, 1520–1605*. Research Paper 201, Department of Geography, University of Chicago.

Lindesay, J. A. and Vogel, C. H. (1990) Historical evidence for Southern Oscillation – southern African rainfall relationships. *International Journal of Climatology* 10, 679–689.

Lipsett-Rivera, S. (1990) Puebla's eighteenth century agrarian decline: a new perspective. *Hispanic American Historical Review* 70 (3): 463–481.

Lipsett Rivera, S. (1992) Indigenous communities and water rights in colonial Puebla: patterns of resistance. *The Americas* 48 (4): 463–483.

Lipsett-Rivera, S. (1993) Water and bureaucracy in colonial Puebla de los Angeles. *Journal of Latin American Studies* 25 (1): 25–44.

Lipsett Rivera, S. (1999a) Agua y supervivencia urbana en el medio rural poblano del siglo XVIII. In Martinez, B. G. and Jacomé, A. G. (eds) *Estudios sobre y ambiente en América Latina, I. Argentina, Bolivia, México, Paraguay*. Mexico City, El Colegio de México/Instituto Panamericano de Geografía e Historia.

Lipsett Rivera, S. (1999b) *To defend our water with the blood of our veins: the struggle for resources in colonial Puebla*. Albuquerque, University of New Mexico Press.

Liu, K., Shen, C. and Louie, K. (2001) A 1,000-year history of typhoon landfalls in Guangdong, Southern China, reconstructed from Chinese historical documentary records. *Annals of the Association of American Geographers* 91 (3): 453–464.

Liverman, D. M. (1990) Drought impacts in Mexico: climate, agriculture, technology, and land tenure in Sonora and Puebla. *Annals of the Association of American Geographers* 80 (1): 49–72.

Liverman, D. M. (1993) The regional impact of global warming in Mexico: uncertainty, vulnerability and response. In Schmandt, J. and Clarkson, J. (eds) *The regions and global warming*. Oxford, Oxford University Press.

Liverman, D. M. (1999) Vulnerability and adaptation to drought in Mexico. *Natural Resources Journal* 39: 99–115.

Liverman, D. M. (2000) Adaptation to drought in Mexico. In Wilhite, D. (ed.) *Drought: a global assessment*, vol. 2 New York, Routledge: 35–45.

Lopez, M. A. D. (2001) The patterns of criminality in the state of Chihuahua: cattle theft in the last decades of the nineteenth century. *Historia Mexicana* 50 (3): 513–553.

Lopez-Lara, R. (1973) El obispado de Michoacán en el siglo XVII. *Estudios Michoacanos* 3. Morelia, Michoacán, Fimax Collección.

Lorenzo, J. L. (1960) Aspectos fisicos del valle de Oaxaca. *Revista Mexicana de Estudios Antropologicos* 16: 49–64.

Lorenzo, J. L. (1987) La etapa Lítica en Mexico. *Antropologia*. Boletín official del Instituto Nacional de Antropología e Historia, nueca epoca, 12. Mexico, Instituto Nacional de Antropología e Historia.

Luers, A. L., Lobell, D. B., Sklar, L., Addams, C. L. and Matson, P. A. (2003) A method for quantifying vulnerability, applied to the agricultural system of the Yaqui Valley, Mexico. *Global Environmental Change* 13: 255–267.

Luterbacher, J., Shmutz, C., Gyalistras, D., Xoplaki, E. and Wanner, H. (1999) Reconstruction of monthly NAO and EI indices back to AD 75. *Geophysical Research Letters* 26: 2745–2748.

Lynch, J. (1965) *Spain under the Hapsburgs*, 2 vols. Oxford, Oxford University Press.

Lynch, J. (1973) *The Spanish American revolutions, 1808–1826*. New York, Norton.

Lynch, J. (1992) The institutional framework of colonial Spanish America. *Journal of Latin American Studies* 24: 69–81.

MacArthur, N. (1970) The demography of primitive populations. *Science* 167: 1097–1101.

McBean, G. (2004) Climate change and extreme weather: a basis for action. *Natural Hazards* 31 (1): 177–190.

McIntosh, R. J., Tainter, J. A. and McIntosh, S. K. (2000) Climate history and human action. In McIntosh, R. J., Tainter, J. A. and McIntosh, S. K. (eds) *The way the wind blows: climate, history and human action*. New York, Columbia University Press: 1–42.

Macleod, M. J. (1973) *Spanish Central America: a socio-economic history, 1520–1720*. Berkeley, University of California Press.

McMichael, A. J., Woodruff, E. E. and Hales, S. (2006) Climate change and human health: present and future risks. *The Lancet* 367: 859–869.

Macnaghton, P. and Urry, J. (1998) *Contested natures*. London, Sage.

McNeill, J. R. (2005) Diamond in the rough: is there a genuine environmental threat to security? A review essay. *International Security* 30 (1): 178–195.

Magaña, V. O. (1999) *Los impactos de El Niño en Mexico*. D. F. Mexico, Dirrección General de Protección Civil, Secretaria de Gobernación.

Magaña, V. O., Conde, C., Sanchez, O. and Gay, C. (1997) Assessment of current and future regional climate scenarios for Mexico. *Climate Research* 9 (1–2): 107–114.

Magaña, V. O., Vázquez, J. L., Pérez, J. L. and Pérez, J. B. (2003) Impact of El Niño on precipitation in México. *Geofísica Internacaional* 42: 313–330.

Malvido, E. (1973) Cronología de las epidemias y crisis agricolas de la epoca colonial. *Historia Mexicana* 89: 96–101.

Manley, G. (1974) Central England temperatures: monthly means 1659 to 1973. *Quarterly Journal of the Royal Meteorological Society* 100: 389–405.

Marcus, J. and Flannery, K. V. (1996) *Zapotec civilization*. London, Thames and Hudson.

Marmolejo, L. (1967) *Efemérides Guanajuatenses o datos para formar la historia de la ciudad de Guanajuato*, Vol. 1. Guanajuato, Universidad de Guanajuato Press.

Marquez-Alameda, A. (ed.) (1992) *Historia General de Chihuahua I: geología, geografía y Arqueología*. Chihuahua, Universidad Autonoma de Ciudad Juarez. Gobierno del Estado de Chihuahua.

Marr, J. S. and Kiracofe, J. B. (2000) Was the Huey Cocoliztli a haemorrhagic fever? *Medical History* 44: 341–362.

Martin, C. E. (1996) *Governance and society in colonial Mexico: Chihuahua in the eighteenth century*. Stanford, Stanford University Press.

Martinez de la Rosa, P. (1965) *Apunte para la historia de Irapuato*. México, Biblioteca de Historia Mexicana, Castalia.

Martinvide, J. and Vallve, M. B. (2005) The use of rogation ceremony records in climatic reconstruction: a case study from Catalunya (Spain). *Climatic Change* 30: 201–221.

Meinzen-Dick, R. and Di Gregorio, M. (2004) Overview. In Meinzen-Dick, R. and Di Gregorio, M. (eds) *Collective action and property rights for sustainable development*. IFPRI 2020 Vision, IFPRI.

Melville, E. G. K. (1990) Environmental and social change in the Valle del Mezquital, Mexico, 1521–1600. *Comparative Studies in Society and History* 32: 24–53.

Melville, E. G. K. (1994) *A plague of sheep: environmental consequences of the Conquest of Mexico*. Cambridge, Cambridge University Press.

Mendoza, B., Jáuregui, E., Diaz-Sandoval, R., García-Acosta, V., Eelasco, V. and Cordero, G. (2005) Historical droughts in central Mexico and their relation with El Niño. *Journal of Applied Meteorology* 44 (5): 709–716.

Messerli, B., Grosjean, M., Hofer, T., Nuñez, L. and Pfister, C. (2000) From nature dominated to human dominated environmental changes, *Quaternary Science Reviews* 19: 459–479.

Metcalfe, S. E. (1987) Historical data and climatic change in Mexico – a review. *Geographical Journal* 153: 211–222.

Metcalfe, S. E., Street-Perrott, F. A., Perrott, F. A. and Harkness, D. D. (1991) Palaeolimnology of the Upper Lerma Basin, Central México: a record of climatic change and anthropogenic distrurbance since 11600 years BP. *Journal of Palaeolimnology* 5: 197–218.

Metcalfe, S. E., Street-Perrott, S. A., O'Hara, S. L., Hales, P. E. and Perrott, R. A. (1994) The paleolimnological record of environmental change: examples from the arid frontier of Mesoamerica. In Millington, A. C. and Pye, K. (eds) *Environmental change in drylands: biogeographical and geomorphological perspectives.* London, Wiley: 131–145.

Meyer, M. C. (1997) *El Agua en el suroest hispanico. Una historia social y legal 1550–1850.* Centro de Investigacion y Estudios Superiores en antropología Social, Mexico.

Meyer, W. B., Butzer, K. W., Downing, T. E., Turner, B. L., (II), Wenzel, G. and Wescoat, J. (1998) Reasoning by analogy. In Raynor, S. and Malone, E. L. (eds) *Human choice and climate change, no. 3: tools for policy analysis.* Columbus, OH, Batelle Press: 218–289.

Michaelowa, A. (2001) The impacts of short-term climate change on British and French agriculture and population in the first half of the eighteenth century. In Jones, P., Ogilvie, A. and Davis, T. (eds) *History and climate: memories of the future?* New York, Kluwer.

Milton, K. (1996) *Environmentalism and cultural theory: exploring the role of anthropology in environmental discourse.* London, Routledge.

Minnis, P. E. (1992) Earliest plant cultivation in the desert borderlands of North America. In Wesley Cohen, C. and Watson, P. J. (eds) *The origins of agriculture: an international perspective.* Washington, DC, Smithsonian Institutional Press: 121–141.

Miranda, J. (1871) Relación hecha por Joan de Miranda, clerigo, al Doctor Orozco, Presidente de la Audiencia de Guadalajara: sobre la tierra y población que hay desde las minas de San Martin a las de Santa Barbara, que esto ultimo entonces estaba poblado, Año de 1575. In *Collección de documentos ineditos relativos al discubrimiento, conquista y organizacion de las antiguas posesiones espanolas de America y Oceania sacados de los archivos del reino,* Vol. 16. Madrid, Imprenta del Hospicio.

Mirza, M. M. Q. (2003) Climate change and extreme weather events: can developing countries adapt? *Climate Policy* 3: 233–248.

Morales, T. and Magaña, V. (1999) Unexpected frosts in central Mexico during summer. *Proceedings of the 10th Symposium on Global Change Studies, 10–15 January, 1999, Dallas, TX.* Preprint volume. Boston, American Meteorological Society: 262–263.

Moreno, H. (1986) *Francisco de Ajofrín, diario del viaje a la Nueva España.* Mexico, Secretaria de Educación Publica.

Moreno, J. W. (1958) La colonizatión y evangelizatión de Guanajuato en el siglo XVI. In *Estudios de Historia Colonial*, Mexico, INAH: 63–94.

Mosiño Alemán, P. and García, E. (1974) The climate of Mexico. In Bryson, R. A. and Hare, K. F. (eds) *Climates of North America*. New York, Elsevier: 345–404.

Mota y Escobar, A. de la (1940) *Descripción geografica de los reinos de Nueva Galicia, Nueva Vizcaya y Nueva León*, 2nd edn. Mexico City, Editorial Pedro Robredo.

Murphy, M. E. (1986) *Irrigation in the Bajio region of colonial Mexico*. Dellplain Latin American Studies, No. 19. Boulder, Westview Press.

Musset, A. (1992) *El agua en al Valle del Mexico: siglos XVI-XVIII*. Mexico, Portico de la Ciudad de Mexico, Centre d'études Mexicanes et Centraméricanes.

Nash, D. J. and Endfield, G. H. (2002) A nineteenth century climate chronology for the Kalahari Desert derived from missionary correspondence. *International Journal of Climatology* 22: 821–841.

Nash, D. J. and Endfield, G. E. (2002b) Missionaries and morals: climatic discourse in nineteenth-century Central Southern Africa. *Annals of the Association of American Geographers* 92 (4): 727–742.

Naylor, T. H. and Polzer, C. W. (1986) *The presidio and militia on the northern frontier of New Spain*. Tuscon, University of Arizona Press.

Neely, J. A. (1972) Prehistoric water supplies and irrigation systems at Monte Alban, Oaxaca, Mexico. Paper presented at the 37th Annual Meeting of the Society for American Archaeology, Miami.

Neely, J. A. and O'Brien, M. J. (1973) Irrigation and settlement nucleation at Monte Alban: a test of models. Paper presented at the 38th Annual Meeting of the Society for American Archaeology, San Francisco.

Neumann, S. I. J. (1969) *Revoltes des Indiens Tarahumars (1626–1724)*. Trans. L. Gonzalez. Paris, Institut des Hautes Etudes de l'Amerique Latine.

Newson, L. A. (1985) Indian population patterns in colonial Spanish America. *Latin American Research Review* 20 (3): 41–74.

Nicholas, L., Feinman, G., Kowalewski, S. A., Blanton, R. E. and Finsten, L. (1986) Prehispanic colonization of the Valley of Oaxaca. *Human Ecology* 14 (2): 131–162.

Nicholson, S. E. (1979) The methodology of historical climatic reconstruction and its application to Africa. *Journal of African History* 20: 31–49.

Nicolle, C. J. (1933) *Destin des maladies infectieuses*. Paris, Librarie Felix Alcan.

NOAA, International Tree Ring Data Bank. ftp://ftp.ngdc.noaa.gov/paleo/treering/.

Nordli, P. Ø. (2001) Reconstruction of nineteenth century summer temperatures in Norway by proxy data from farmers' diaries. *Climatic Change* 48: 201–218.

Nuñez Cabeza de Vaca (1958) '*Naufragíos*' en *Viajes y viajeros viajes por Norteamérica*. Madrid Aguilar, Biblioteca Indiana: 17–59.

O'Brien, K. L. and Leichenko, R. M. (2000) Double exposure: assessing the impacts of climate change within the context of globalization. *Global Environmental Change* 10: 221–232.

O'Brien, K. L. and Leichenko, R. M. (2003) Winners and losers in the context of global change. *Annals of the Association of American Geographers* 93: 89–103.

O'Brien, M. J., Lewarch, D. E., Mason, R. D. and Neely, J. A. (1980) Functional analysis of water control features at Monte Alban, Oaxaca, Mexico. *World Archaeology* 11 (3): 342–355.

O'Hara, S. L. (1993) Historical evidence of fluctuations in the level of Lake Pátzcuaro, Michoacán, Mexico over the last 600 years. *Geographical Journal* 159: 51–62.

O'Hara, S. L. and Metcalfe, S. E. (1995) Reconstructing the climate of Mexico from historical records. *The Holocene* 5 (4): 485–490.

Ohlsson, L. (2000) *Livelihood conflicts – linking poverty and environment as causes of conflict.* Stockholm, Sida, Environmental Policy Unit.

Oliver-Smith, A. and Hoffman, S. M. (eds) (1999) *The angry earth: disasters in anthropological perspective.* London, Routledge.

Orozco y Berra, M. (1938) *Historia de la dominación española en México, con una advertencia de Genaro estrada*, 2 vols. Mexico, Biblioteca Historica Mexicana de obras ineditas Núm 10: 242–248.

Ortlieb, L. (1999) The documented historical record of El Niño events in Peru: an update of the Quinn record (sixteenth through nineteenth centuries). In Diaz, H. F. and Markgraf, V. (eds) *El Niño and the Southern Oscillation: multiscale variability and global and regional impacts.* Cambridge, Cambridge University Press: 207–295.

Ouweneel, A. (1991) Growth, stagnation, and migration: an explorative analysis of the tributario series of Anahuac (1720–1800). *Hispanic American Historical Review* 7: 531–577.

Ouweneel, A. (1996) *Shadows over Anáhuac: an ecological interpretation of crisis and development in central México 1730–1800.* Albuquerque, University of New Mexico Press.

Paavola, J. and Adger, W. N. (2006) Fair adaptation to climate change. *Ecological Economics* 56: 594–609.

Padilla, G., Rordriguez, L., Castorena, G. and Florescano, E. (1980) *Análisis historico de las sequias en Mexico.* Mexico, SARH.

Palerm, A. and Wolf, E. R. (1957) *Ecological potential and cultural development in Mesoamerica.* Panamerican Union Social Science Monograph 3: 1–37.

Paredes, C. M. (1979) *El sistema tributario prehispánico entre los Tarascos.* Unpublished manuscript.

Parker, G. and Smith, L. M. (1997) Introduction. In Parker, G. and Smith, L. M. (eds) *The general crisis of the seventeenth century.* London, Routledge: 1–55.

Parmenter, R. R., Yadav, E. P., Parmenter, C. A., Ettestad, P. and Gage, K. L. (1999) Incidence of plague associated with increased winter-spring precipitation in New Mexico. *American Journal of Tropical Medicine and Hygiene* 61 (5): 814–821.

Parry, M. (2001) Viewpoint: climate change: where should our priorities be? *Global Environmental Change* 11: 257–260.

Pascual, M., Rodó, X., Ellner, S. P., Colwell, R. and Bouma, M. J. (2000) Cholera dynamics and El Niño-Southern Oscillation. *Science* 289 (5485): 1766–1769.

Paso y Troncoso, F. (1939–40) *Papeles de la Nueva España*, 2a serie. Mexico.

Pastor, R. (1987) El repartimiento de mercancias y los alcades mayores novohispanos. Un sistema de exploitacion de su origen a la crisis de 1810. In Borah, W. (ed.) *El Gobierno provincial en Nueva España, 1570–1787.* Mexico, UNAM: 201–236.

Patz, J. A., Campbell-Lendrum, D., Holloway, T. and Foley, J. A. (2005) Impact of regional climate change on human health. *Nature* 438 (17): 310–317.

Patz, J. A., McGeehin, M. A., Bernard, S. M., Ebi, K. L., Epstein, P. R., Grambsch, A., Gubler, D. J., Reiter, P., Romieum, L., Rose, J. B., Samet, J. M. and Trtanj, J.

(2000) The potential health impacts of climate variability and change for the United States: executive summary of the report of the health sector of the US National Assessment. *Environmental Health Perspectives* 109: 367–376.

Pellicer, S. N. (1994) Las transformaciones de la económia indígena en Michoacán: siglo XVI. In Rojas-Rabiela, T. (ed.) *Agricultura indigena pasado y presente.* Mexico Ediciones de la Casa Chata. Centro de Investigaciones y Estudios Superiores en Antropología: 109–128.

Pelling, M. and High, C. (2005) Understanding adaptation: what can social capital offer assessments of adaptive capacity? *Global Environmental Change* 15: 308–319.

Perre, J. (1970) La peste de 1643 en Michoacán. In *Historia y Sociedad en el Nuevo Mundo de habla española, Homenaje a José Miranda.* México, El Colegio de México: 247–261.

Peterson, J. F. (1992) The Virgin of Guadelupe: symbol of conquest of liberation. *Art Journal* 51 (4): Latin American Art: 39–47.

Pfister, C. (2005) Weeping in the snow: the second period of Little Ice Age-type impacts, 1570 to 1630. In Behringer, W., Lehmann, H. and Pfister, C. (eds.) *Kulturelle Konsequenzen der Kleinen Eiszeit – Cultural consequences of the Little Ice Age*, Göttingen, Vandenhoek and Ruprecht.

Pfister, C. (in preparation) Natural disasters – catalysts for fundamental learning. In Pfister, C. and Mauch, C. (eds) *Natural hazards: cultural responses in global perspective.* Letington Books.

Pfister, C. and Brazdil, R. (1999) Climatic variability in sixteenth century Europe and its social dimension: a synthesis. *Climatic Change* 43: 5–53.

Pfister, C., Brázdil, R. and Glaser, R. (eds) (1999) *Climatic variability in sixteenth-century Europe and its social dimension.* Dordrecht, Kluwer.

Pfister, C., Brazdil, R., Obrebska-Starkel, B., Starkel, L., Heino, R. and von Storch, H. (2001) Strides made in reconstructing past weather and climate. *Eos – Transactions of the American Geophysical Union* 82: 248.

Piervitali, E. and Colacino, M. (2001) Evidence of drought in Western Sicily during the period 1565–1915 from liturgical offices. *Climatic Change* 49 (1–2): 225–238.

Pietschmann, H. (1988) Agricultura e industria rural indígena en el Mexico de la segunda mitad del siglo XVIII. In Ouweneel, A. and Torales Pacheco, C. (eds) *Empresarios indios y estado: perfil de la economía Mexicana* (Siglo XVIII). Amsterdam, CEDLA.

Ponzio, C. A. (2005) Globalization and economic growth in the third world: some evidence from eighteenth century Mexico. *Journal of Latin American Studies* 37: 437–467.

Portillo, A. (1910) *Oaxaca en el centenario de la Independencia nacional.* Oaxaca.

Powell, P. W. (1944) Presidios and towns on the silver frontier of New Spain, 1550–1580. *Hispanic American Historical Review* 24 (2): 179–200.

Powell, P. W. (1945) The Chichimecas: scourge of the silver frontier in sixteenth century Mexico. *Hispanic American Historical Review* 25 (3): 315–338.

Prem, H. J. (1974) El Rio Cotzala: estudio historico de un sistema de riego. *Comunicaciones* 11: 53–67.

Prem, H. J. (1984) Early Spanish colonization and Indians in the Valley del Atlixo, Puebla. In Harvey, H. R. and Prem, H. J. (eds) *Explorations in ethnohistory: Indians in central Mexico in the sixteenth century.* Albuquerque, University of New Mexico Press: 205–228.

Prem, H. J. (1992) Spanish colonization and Indian property in central Mexico, 1521–1620. *Annals of the Association of American Geographers* 82 (3): 444–461.

Prieto, M. del R. (1983) El clima de Mendoza durante los siglos XVII y XVIII. *Meteorologica* 14 (102): 129–138.

Prieto M. del R., García-Herrera, R. and Dussel, P. (2000) Archival evidence for some aspects of historical climate variability in Argentina and Bolivia during the 17th and 18th centuries. In Volkheimer, W. and Smolka, P. (eds) *Southern hemisphere palaeo and neoclimates*. Berlin, Springer.

Prieto, M. del R., García-Herrera, R. and Hernández Martin, E. (2004) Early records of icebergs in the South Atlantic Ocean from Spanish documentary sources. *Climatic Change* 66 (1–2): 29–48.

Quinn, W. H. and Neal, V. T. (1992) The historical record of El Niño events. In Bradley, R. S. and Jones, P. D. (eds) *Climate since AD 1500*. London, Routledge: 623–648.

Rabel Romero, C. E. (1975) San Luis de la Paz, estudio de economica y demografía historicas, 1645–1810. Ph.D. diss., Institute Nacional de Antropología y Historia, Mexico.

Recopiliación (1987) *Recopiliación de leyes de los Reynos de las Indias* (1681). Mexico City, Porrúa (facsimile edition).

Retsö, D. (2002) A contribution to the history of European winters: some climatological proxy data from early-sixteenth century Swedish documentary sources. *Climatic Change* 52: 137–173.

Ribot, J. C. (1995) The casual structure of vulnerability: its application to climate impacts analysis. *Geo-Journal* 35: 119–122.

Riley, C. L. (1976) Las Casas and the golden cities. *Ethnohistory*: 19–30.

Riley, J. D. (2002) Public works and local elites: the politics of taxation in Tlaxcala 1780–1810. *The Americas* 58 (3): 355–393.

Roberts, L. (1989) Disease and death in the New World. *Science* 246: 1245–1247.

Rocha, J. (1942) Datos sobre la fundación de Ciudad Jiminez. *Boletín de la Sociedad Chihuahuense de Estudios Historicos* 4.

Rodrigo, F. S., Esteban-Parra, M. J. and Castro-Diez, Y. (1998) On the use of the Jesuit order private correspondence records in climate reconstructions: a case study from Castille (Spain) for AD 1634–1648. *Climatic Change* 40: 625–645.

Rodrigues, M. S. (2005) Uso de cajas de agua. In Grijalva, M. N. (ed.) *Los usos de agua en el centro y norte de Mexico. Historiografia, technologia, conflictos.* Zacatecas, Mexico, Universidad Autonoma de Zacatecas Press.

Rodriguez, G. L. (1987) *Cronicas de la Sierra Tarahumara*. Mexico, SEP.

Ropelewski, C. F. and Halpert, M. S. (1987) Global and regional scale precipitation patterns associated with the El Niño Southern Oscillation. *Monthly Weather Review* 115: 1606–1626.

Ropelewski, C. F. and Halpert, M. S. (1989) Precipitation patterns associated with the high index phase of the Southern Oscillation. *Journal of Climatology* 1: 268–284.

Rosenblat, A. (1992) The population of Hispaniola at the time of Columbus. In Denevan, W. M. (ed.) *The Native Population of the Americas in 1492*. Madison, University of Wisconsin Press: 43–66.

Ruiz, R. E. (1992) *Triumphs and tragedy: a history of the Mexican people*. New York, W. W. Norton.

Salmon, R. M. (1977) Tarahumara resistance to mission congregation in northern New Spain, 1580–1710. *Ethnohistory* 24 (4): 379–393.

Sanders, W. T. (1992) Ecology and cultural syncretism in sixteenth century Mesoamerica. *Antiquity* 66: 172–190.

Sanders, W. T. and Nichols, D. L. (1988) Ecological theory and cultural evolution in the Valley of Oaxaca. *Current Anthropology* 29 (1): 33–80.

Sandoval, C. A. E. (2003) Vulnerability due to extremes: droughts. Enzo Levi Lecture, 2002. *Ingeneria Hidraulica en Mexico* 18 (2): 133–155.

Santiago, M. (1996) Eighteenth century military policy in northern New Spain. *Journal of Arizona History* 37: 283–290.

Saunders, P. L. (2000) Environmental refugees: the origins of a construct. In Stott, P. and Sullivan, S. (eds) *Political ecology: science, myth and power.* London, Arnold: 218–246.

Schmidt, R. H. (1992) Chihuahua, tierra de contrastes geograficos. In Lau, R. (ed.) *Historia general de Chihuahua I. Geologia, geográfia y arqueología.* Universidad Autonoma de Ciudad Juarez. Gobierno del Estado de Chihuahua: 47–101.

Scott, J. (1976) *The moral economy of the peasant: rebellion and subsistence in Southeast Asia.* New Haven, Yale University Press.

Scott, J. (1985) *Weapons of the weak: everyday forms of peasant resistance.* New Haven, Yale University Press.

Seielstad, G. A., Shea, E., Vogel, C. and Wilbanks, T. J. (2000) Assessing vulnerability to global environmental risks. Report of the Workshop on Vulnerability to Global Environmental Change: Challenges for Research, Assessment and Decision Making.

Serulnikov, S. (1996) Disputed images of colonialism: Spanish rule and Indian subversion in northern Potosi. *Hispanic American Historical Review* 76 (2): 189–226.

Servin, C. C. (2005) Las sequias en Mexico dirante el siglo XIX. Investigaciones Geográficas. *Boletin del Instituto de Geográfica,* UNAM.

Simpson, L. B. (1952) *Exploitation of land in central Mexico in the sixteenth century.* Berkeley, University of California Press.

Simpson, L. B. (1966) *Many Mexicos.* Berkeley, University of California Press.

Slicher van Bath, B. H. (1978) The calculation of the population of New Spain, especially for the period before 1570. *Boletin de Estudios Latinoamerica y del Caribe* 24: 67–95.

Sluyter, A. (2001) Colonialism and landscape in the Americas: material/conceptual transformations and continuing consequences. *Annals of the Association of American Geographers* 9: 410–428.

Sluyter, A. (2002) *Colonialism and landscape: postcolonial theory and applications.* New York, Rowman and Littlefield.

Smiley, F. E. (1994) The agricultural transitions in the northern southwest: patterns in the current chronometric data. *The Kiva* 60: 165–189.

Smit, B., Burton, I., Kelin, R. J. T. and Wandel, J. (2000) An anatomy of adaptation to climate change and variability. *Climatic Change* 45: 223–251.

Smithers, J. and Smit, B. (1997) Agricultural system response to environmental stress. In Ilbery, B., Chiotti, O. and Rickard, T. (eds) *Agricultural restructuring and sustainability.* Wallingford, CAB International: 167–183.

Smout, T. C. (1979) *A history of the Scottish people, 1560–1830.* Glasgow, Collins.

Solórzano y Pereyra, J. de (Compilador) (1776) *Política Indiana*. Madrid, Real de la Gazeta.

Sousa, L. and Terraciano, K. (2003) The "Original Conquest" of Oaxaca: Nahua and Mixtec accounts of the Spanish conquest. *Ethnohistory* 50 (2): 349–400.

Spores, R. (1967) *The Mixtec kings and their people*. Norman, University of Oklahoma Press.

Spores, R. (1972) *An archaeological settlement survey of the Nochixtlan Valley, Oaxaca*. Vanderbilt University Publications in Anthropology 11. Nashville, Vanderbilt University.

Stahle, D. W., Villanueva-Diaz, J., Cleaveland, M. K., Therrell, M., Paul, G. J., Burns, B. T., Salinas, W., Suzan, H. and Fule, P. Z. (2000) Recent tree-ring research in Mexico. In Roig, F. A. (compiler) *Dendrochronología en America Latina*. Mendoza, Argentina, Editorial de La Universadad nacional de Cuyo: 285–305.

Stein, S. J. (1981) Bureaucracy and business in the Spanish empire, 1759–1804: failure of a Bourbon reform in Mexico and Peru. *Hispanic American Historical Review* 61 (1): 2–28.

Stein, S. J. and Stein, B. H. (1970) *The colonial heritage of Latin America: essays on economic dependence in historical perspective*. New York, Oxford University Press.

Stothers, R. (1996) The great dry fog of 1783. *Climatic Change* 32: 79–89.

Streets, D. G. and Glantz, M. H. (2000) Exploring the concept of climate surprise. *Global Environmental Change – Human and Policy Dimensions* 10 (2): 97–107.

Swan, S. C. (1981) Mexico in the Little Ice Age. *Journal of Interdisciplinary History* 14 (4): 633–648.

Swetnam, T. W., Allen, C. D. and Betancourt, J. L. (1999) Applied historical ecology: using the past to manage for the future. *Ecological Applications* 9 (4): 1189–1206.

Tamarón y Romeral, P. (1937) *Demonstración del vastisimo obispado de la Nueva Vizcaya*. Biblioteca Mexicana de Obras Ineditas, 7, 1765, Mexico City, Antigua Libreria Robredo de Jose Purrua e Hijos.

Taylor, W. B. (1972) *Landlord and peasant in colonial Oaxaca*. Stanford, Stanford University Press.

Taylor, W. B. (1978) Haciendas coloniales en el Valle de Oaxaca. In Florescano, E. (coord) *Haciendas, latifundios y plantaciones en America Latina*: 74.

Taylor, W. B. (1987) The Virgin of Guadelupe in New Spain: an inquiry into the social history of Marian devotion. *American Ethnologist* 14 (1): 9–33.

TePaske, J. T. and Klein, H. S. (1981) The seventeenth century crisis in New Spain: myth or reality? *Past and Present* 90: 116–135.

Therrell, M. D., Stahle, D. W. and Acuña-Soto, R. (2004) Aztec drought and the 'curse of one rabbit'. *Bulletin of the American Meteorological Society* (September): 1263–1272.

Therrell, M. D., Stahle, D. W., Villanueva Diaz, J., Cornejo Oviedo, E. H. and Cleaveland, M. K. (2006) Tree ring reconstructed maize yield in central Mexico, 1474–2001. *Climatic Change* 74: 493–504.

Thompson, I. A. A. and Yun Casalilla (eds) (1994) *The Castilian crisis of the seventeenth century*. Cambridge, Cambridge University Press.

Tilly, C. (1996) Conclusion: contention and the urban poor in eighteenth and nineteenth century Latin America. In Arrom, S. M. and Ortoll, S. (eds) *Riots in the cities: popular politics and the urban poor in Latin America, 1765–1910.* Wilmington, Scholarly Resources: 225–242.

Tompkins, E. L. and Adger, W. N. (2004) Does adaptive management of natural resources enhance resilience to climate change? *Ecology and Society* 9 (2): 10. www.ecologyandsociety.org/vol9/iss2/art10.

Turner, B. L., Matson, P. A., McCarthy, J. J., Corell, R. W., Christensen, L., Eckley, N., Hovelsrud-Broda, G. K., Kasperson, J. X., Kasperson, R. E., Luers, A., Martello, M. L., Mathiesen, S., Naylor, R., Polsky, C., Pulsipher, A., Schiller, A., Selin, H. and Tyler, N. (2003) Illustrating the coupled human-environment system for vulnerability analysis: three case studies. *Proceedings of the National Academy of Sciences* 100: 8080–8085.

Tutino, J. (1986) *From insurrection to revolution in Mexico.* Princeton, Princeton University Press.

Tutino, J. (1998) The revolution in Mexican independence: insurgency and the renegotiation of property, production and patriarchy in the Bajío, 1800–1855. *Hispanic American Historical Review* 78 (3): 367–418.

Tweedie, M. J. (1968) Notes on the history and adaptation of the Apache tribes. *American Anthropology* 70: 1132–1142.

Ugarte, B. S. J. J. (1992) *Historia sucinta de Michoacán.* Morela, Michoacán, Morevallado Editores.

Van Young, E. (1981) *Hacienda and market in eighteenth-century Mexico: the rural economy of the Guadalajara Region, 1675–1820.* Berkeley, University of California Press.

Van Young, E. (1986) Millennium on the northern marches: the Mad Messiah of Durango and popular rebellion in Mexico, 1800–1815. *Comparative Studies in Society and History* 28 (3): 385–413.

Van Young, E. (1988) Islands in the storm: quiet cities and violent countryside in the Mexican independence era. *Past and Present* 118: 130–155.

Van Young, E. (1993) Agrarian rebellion and defense of community: meaning and collective violence in late colonial and independence era Mexico. *Journal of Social History* (Winter): 245–269.

Vermillion, D. L. (2001) Property rights and collective action in the devolution of irrigation system management. In Meinzen-Dick, R., Knox, A. and Di Gregorio, M. (eds.) *Collective action, property rights and devolution of natural resource management: exchange of knowledge and implications for policy.* Germany, DSE/ZEL: 183–220.

Vierra, B. J. (1994) Archaic hunter-gatherer mobility strategies in Northwestern New Mexico. In Vierra, B. J. (ed.) *Archaic hunter-gatherer archaeology in the American Southwest, Contributions in Anthropology* 13 (1). Portales, Eastern New Mexico University Press: 121–154.

Villamarín, J. and Villamarín, J. (1992) Epidemic disease in the Sabana de Bogota, 1536–1810. In Cook, N. D. and Lovell, G. W. (eds) *Secret judgments of God: old world disease in colonial Spanish America.* Norman, University of Oklahoma Press: 113–141.

Viruell, L. A. A. D. (2004) La producción de grana cochinilla en Oaxaca a principios del siglo XIX. *Boletín del Archivo General de la Nación* 5: 35–66.

Vogel, C. (1989) A documentary-derived climatic chronology for southern Africa, 1820–1900. *Climatic Change* 14: 291–306.

Vogel, C. (2001) *Vulnerability and global environmental change.* Draft paper for the Human Dimensions of Global Change Meeting, Rio, October.

Wallen, C. (1955) Some characteristics of precipitation in Mexico. *Geofisika Annaler* 37: 51–85.

Warrenton, V. A. Research and Assessment Systems for Sustainability Program Discussion Paper 2000–12. Environment and Natural Resources Program, Belfer Center for Science and International Affairs (BCSIA). Cambridge, MA, Kennedy School of Government, Harvard University.

Webre, S. (1990) Water and society in a Spanish American city: Santiago de Guatemala. *Hispanic American Historical Review* 70 (1): 57–84.

West, R. C. (1949) The mining community in Northern New Spain: the Parral mining district. *Iberoamericana* 30. Berkeley, University of California Press.

Whitmore, T. M. (1991) A simulation of sixteenth century population collapse in the Basin of Mexico. *Annals of the Association of American Geographers* 81 (4): 464–487.

Whitmore, T. M. (1992) *Disease and death in early colonial Mexico: simulating Amerindian depopulation.* Dellplain Latin American Geography Series, No. 28. Boulder, Westview Press.

Whitmore, T. M. and Turner, B. L. (1992) Landscapes of cultivation in Mesoamerica on the eve of Conquest. *Annals of the Association of American Geographers* 83: 402–425.

Wigley, T. M. L. (1985) Impact of extreme events. *Nature* 316: 106–107.

Wigley, T. M. L., Ingram, M. J. and Farmer, G. (1981) *Climate and history.* Cambridge, Cambridge University Press.

Wilken, G. C. (1987) *Good farmers: traditional agricultural resource management in Mexico and Central America.* Berkeley, University of California Press.

Williams, B. (1972) Tepetate in the Valley of Mexico. *Annals of the Association of American Geographers* 62: 618–626.

Winter, M. C. (1989) *Oaxaca: the archaeological record.* Mexico City, Minutiae Mexicana.

Witham, C. S. and Oppenheimer, C. (2005) Mortality in England during the 1783–4 Laki Craters eruption. *Bulletin of Volcanology* 67 (1): 15–26.

Wolf, E. (1953) La formación de la nación: un ensayo de formulación. *Ciencias Sociales* 4 (20, 21, 22). Oficina de Ciencias Sociales de la Union Panamericana.

Wright Carr, D. C. (1989) *Querétaro en el siglo XVI, fuentes documentales primarias.* Querétaro, Secretaria de Cultura y Bienestar Social, Gobierno del Estado de Querétaro.

Wright Carr, D. C. (1998) *La conquista del Bajío y los orígenes de San Miguel de Allende.* Mexico, FCE y UVM.

Wrightson, K. (1989) Kindred adjoining kingdoms: an English perspective on the social and economic history of early modern Scotland. In Houston, R. A. and Whyte, R. D. (eds) *Scottish history 1500–1800.* Cambridge, Cambridge University Press, 245–260.

Yates, P. L. (1981) *Mexico's agricultural dilemma.* Tucson, University of Arizona Press.

Yin, H. and Li, C. (2001) Human impact on floods and flood disasters on the Yangtze River. *Geomorphology* 41 (2): 105–109.

Zamarroni Arroyo, R. (1960) *Naraciones y Leyendas de Celaya y del Bajío*, 2 vols. Mexico, Editorial Periodistica e Impresora de Mexico.

Zeigler, D. J., Brunn, S. D. and Johnson, J. H. (1996) Focusing on Hurricane Andrew through the eyes of the victims. *Area* 28 (2): 124–129.

Zeitlin, R. N. (1990) The Isthmus and the Valley of Oaxaca: questions of Zapotec imperialism in Formative period Mesoamerica. *American Antiquity* 55 (2): 250–261.

Zell, R. (2004) Global climate change and the emergence/re-emergence of infectious diseases. *International Journal of Medical Microbiology* 293, Supplement 37: 16–26.

Zhang, D. (1994) Evidence for the existence of the medieval warm period in China. *Climatic Change* 26 (2–3): 289–297.

Zinsser, H. (1935) *Rats, lice and history*. Boston, Little Brown.

Index